Electrodialysis
http://dx.doi.org/10.5772/intechopen.87853
Edited by Taner Yonar

Contributors
Noureddine Toujeni, Birsen Cevher Keskin, Jonathan Deseure, Jérôme Aicart, Biman Gati Gupta, Juvenal Rocha Dias, Abril Gardenia Martínez Castillo, Jorge Octaviano Gomez Castrejon, Juan Carlos Gómez Buendía, Jorge A. Mendoza-Pérez, Taner Yonar, Nahla Bouaziz, Lakder Kairouani, Eliane Aparecida Faria Amaral Fadigas

First published in London, United Kingdom, 2020 by IntechOpen
IntechOpen is the global imprint of INTECHOPEN LIMITED, registered in England and Wales, registration number: 11086078, 5 Princes Gate Court, London, SW7 2QJ, United Kingdom
Printed in Croatia

British Library Cataloguing-in-Publication Data
A catalogue record for this book is available from the British Library

Additional hard and PDF copies can be obtained from orders@intechopen.com

Electrodialysis
Edited by Taner Yonar
p. cm.
Print ISBN 978-1-83968-382-4
Online ISBN 978-1-83968-383-1
eBook (PDF) ISBN 978-1-83968-384-8

For EU product safety concerns:
IN TECH d.o.o., Prolaz Marije Krucifikse Kozulić 3, 51000 Rijeka, Croatia, info@intechopen.com or visit our website at intechopen.com.

Contents

Preface

The electrodialysis process, from energy to the environment, offers very important opportunities. Separation of ion content in water, recovery, electrolysis of water, and many other possibilities will be possible with electrodialysis. Electrodialysis is of great importance not only in the efficient use of resources in our world, but also in space studies. In this book on electrodialysis, in 7 chapters, the possibilities offered by the electrodialysis process are discussed in detail and presented to the service of the scientific community. In the first chapter, Toujani et al. discussed the organic rankine cycle (ORC) application on desalination systems. In the second chapter, Cevher-Keskin describe how the functional organization of eukaryotic cells requires the exchange of proteins, lipids, and polysaccharides between membrane compartments through transport intermediates. The vacuolar and endocytic trafficking are presented to enhance our understanding of plant development and immunity in plants. In the third chapter, the breakdown of endocrine disrupting drugs (trimethoprim and a mixture of clavulanic acid-amoxicilli) in aqueous media using electrolysis by Perez et al is discussed. The fourth chapter proposes an alternative system for conventional reverse osmosis (RO) and electrodialysis (ED) desalination plants by incorporating the use of gravitational potential energy (GPE). The proposed system in this chapter is devised with two subsystems, the RO module followed by the ED module, both simultaneously powered by GPE. The fifth chapter proposes a zero discharge approach for the sustainable protection of groundwater by membrane-based treatment processes by Gupta. In the sixth chapter, Deseure and Aicart provide the details of hydrogen production by Solid Oxide Steam Electrolyser (SOEC) coupled with renewable energy sources. Finally, in the seventh chapter, Yonar briefly discusses the potential of the Electrodialysis Bipolar Membrane Process potential on membrane concentrate management.

Taner Yonar
Environmental Engineering Department,
Bursa Uludag University,
Bursa, Turkey

Chapter 1

Performance Analysis of a New Combined Organic Rankine Cycle and Vapor Compression Cogeneration and Tri-Generation and Water Desalination

Noureddine Toujani, Nahla Bouaziz and Lakder Kairouani

Abstract

The new ORC-VCC combined system is analyzed. It is a new system that can be operated in four modes depending on the type of energy. The novelty of the system appears essentially in the development of new ORC-VCC combination architecture, the lowering of the condensation temperature, the possibility of cold production by the ORC cycle affected by the pumping phase, preheating of fluid cycle using the VCC cycle fluid, and new configurations based on the integration of heat recovery systems to improve overall system performance. In addition, each installation mode has several configurations depending on the recovery points that will be integrated later, besides its adaptation to any energy source, where we can use biomass, solar, and heat rejects of industry at low temperatures (60–130°C). This system can produce under and above zero temperature. Although, due to its architecture, it is also characterized by many combination of selection fluid for the ORC and VCC cycles, it is not necessary to have the same working fluid as in the classic systems. In this study, three configurations are examined and studied in terms of energy efficiency mainly the performance of each configuration including net power, refrigeration capacity and overall efficiency, the thermal efficiency for ORC.

Keywords: organic Rankine cycle, tri-generation, vapor compression cycle, desalination of seawater

1. Introduction

According to the International Energy Agency (IEA), solar power will be the fastest-growing source of energy in the future. The growth rate of solar energy can reach more than 12% [1]. Many countries today are making decisions to put political strategies in the use of renewable resources. For that, many studies were done all over the world, Asia [2, 3], Africa [4, 5], and America [6], whose objectives are to determine the energy potential and to choose the political strategies to improve the solar energy potential. In Europe [1], the Commission communication to the European Parliament and the Council for new European energy policies set out in

2014 [7] a target to reach 20% energy efficiency by 2020 and 30% by 2030. In fact, researches are looking for technologies that can be used to mitigate global warming around the world and reduce CO_2 emissions [8]. Energy shortage problems are faced in all over the world and becoming more acute in all fields [9, 10] (metallurgy, chemical, electrical and mechanical sectors). So, the world is facing two energy challenges: increase production to meet energy needs and reduce CO_2 emissions issued by industrial plants. For that, the utilization of renewable energy becomes a political duty and not a strategic choice to solve energy problems. Among these renewable resources used is the solar energy: Sunil Kumar [11] published a review as a synthetic fruit of studies done on energy analysis. He presented the various solar energy systems used in solar drying [12, 13], solar air conditioning [14, 15], solar refrigeration [16], solar water heating [17], and solar cooking [18]. These systems have been operated by solar photovoltaic techniques [19] and solar thermal energy used for heat and power generation [20–22].

In terms of management and tri-generation system design, several studies [23–28] analyzed the energy potential through the integration and hybridization of renewable sources.

The new ORC-VCC combined system is developed. As it is shown in **Figures 1–4**, we wanted to design a new architecture for multi-objective optimization. It is a new system that can be operated in four modes depending on the type of the produced energy, namely, the electric energy, refrigeration, poly-generation, and water desalination. The four developed modes are:

- **Mode 1**: cold production. **Figure 1** illustrates the basic architecture of the system. It receives a heat flow from an external renewable source in the boiler so that the ORC cycle can be run in order to deliver a mechanical work at the turbine; this work is transmitted totally to the VCC cycle compressor (turbo system compressor). This system provides us a refrigeration quantity at the evaporator as illustrated in the figure.

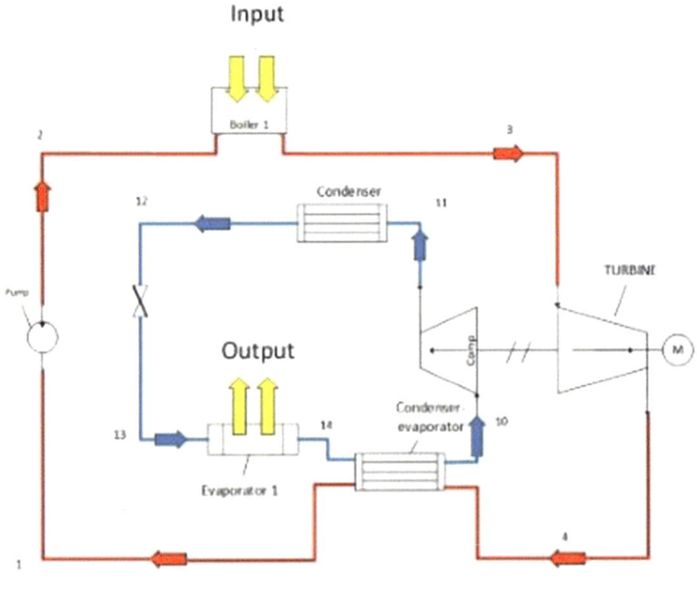

Figure 1.
Cold production mode.

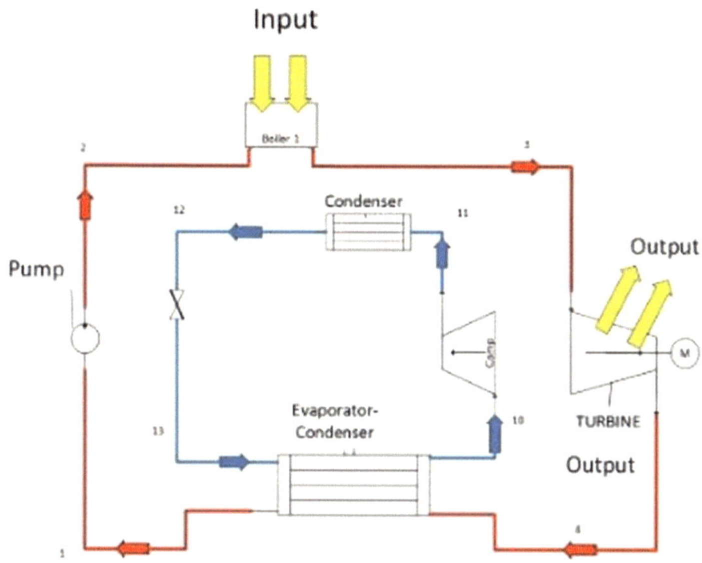

Figure 2.
Electricity production mode.

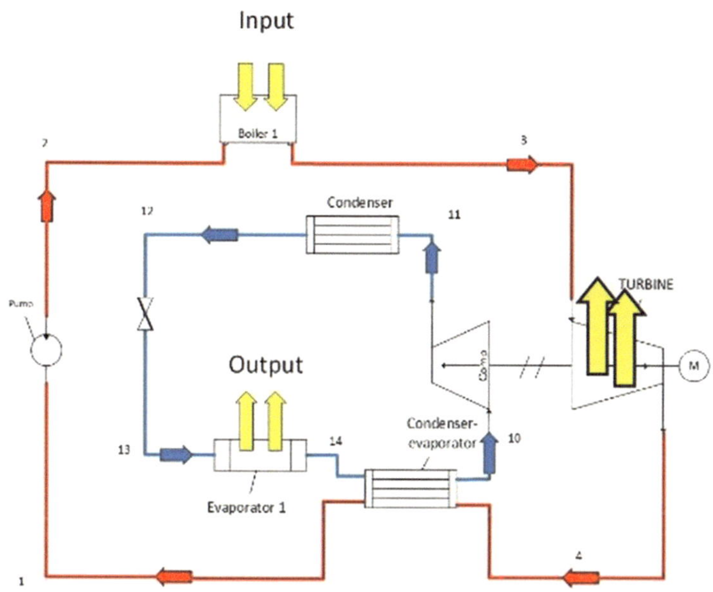

Figure 3.
Cogeneration production mode.

- **Mode 2**: electricity power. **Figure 2** shows the basic installation. It also receives a quantity of heat from an external renewable source in the boiler to have mechanical work at the turbine; it is partially transmitted to the VCC cycle compressor. On the other hand, the power supplied by the VCC cycle evaporator is totally exploited by the ORC cycle condenser. So this mode of operation requires a renewable source and provides us an electric power.

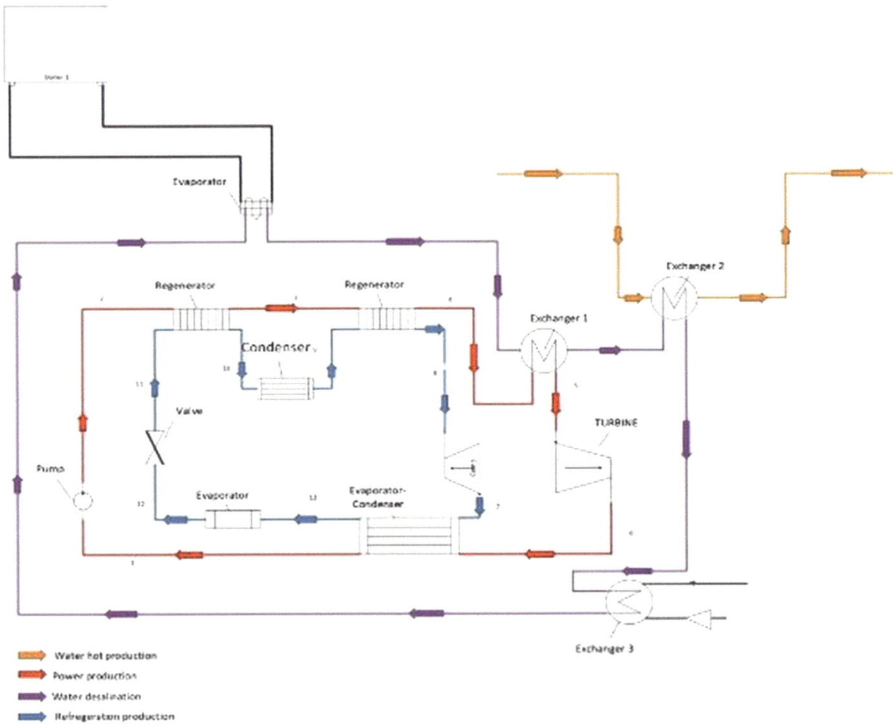

Figure 4.
Tri-generation and desalination mode.

- **Mode 3**: cogeneration (cold production and electricity power). **Figure 3** presents the basic architecture. It receives an external renewable source in the boiler. Through this source, it allows us to have mechanical work at the turbine: this is partially transmitted to the VCC cycle compressor as mode 2. The power provided by the VCC cycle evaporator is partially operated by the condenser ORC cycle. So this operation mode requires a renewable source and offers an electric power by ORC cycle and cooling capacity by the VCC cycle.

- **Mode 4**: tri-generation and desalination of seawater are illustrated in **Figure 4**. This configuration has four circuits: an ORC cycle circuit that is represented by the red color, a circuit of the VCC cycle in blue, a circuit in purple color of the desalinated seawater, and a red circuit of the heated water. We will couple the system with a limited renewable energy source.

In addition, each installation mode has several configurations depending on the recovery points that will be integrated later, besides its adaptation to any energy source, where we can use biomass, solar, and heat rejects of industry at low temperatures (60–130°C). This system could produce a negative and a positive cold. Although, due to its architecture, it is also characterized by many combinations of selection fluid for the ORC and VCC cycles, it is not necessary to have the same working fluid as the classic systems.

The main purpose of this presented study is to analyze the performance of a new system that combines the steam compression cycle and the Rankine cycle for tri-generation (electricity, cold, hot) as well as the desalination of water. This system uses a low-temperature heat source such as solar energy, heat from industrial waste, and biomass.

The objectives of this study are:

- Architectural development of the basic system
- Development of improvement configurations
- Energy analysis and choice of fluids
- The impact of operating parameters on energy performance

In this study, we will develop a new ORC combination with the VCC system in order to make cogeneration and tri-generation with a negative temperature cold (−10°C, 0°C), as well with a positive temperature cold (0°, 10°C). Three new configurations are examined and studied in terms of energy efficiency, namely, the performance of each configuration including net power, refrigeration capacity and overall efficiency, the thermal efficiency for ORC, and the coefficient of performance for VCC. The working fluids are n-hexane for ORC and R600 for VCC.

2. System description

As illustrated in **Figure 5**, the configuration consists of four circuits: an ORC cycle circuit represented by the red color, a circuit of the VCC cycle which is in blue, a circuit in purple color of the desalinated seawater, and a red circuit in the heated water. We will couple our facility with a limited renewable energy source which is thermal photovoltaic center, at low temperature (100–130°C). Our approach is to lower the condensing temperature between −10 and 10°C of the ORC cycle so that the delivered work can be increased. So, a cold part produced by VCC will be dedicated to condense the fluid of the ORC cycle. For this, we will integrate an exchanger regenerator1 which is used to condense the ORC fluid by a quantity of cold produced laying vapor phase VCC side.

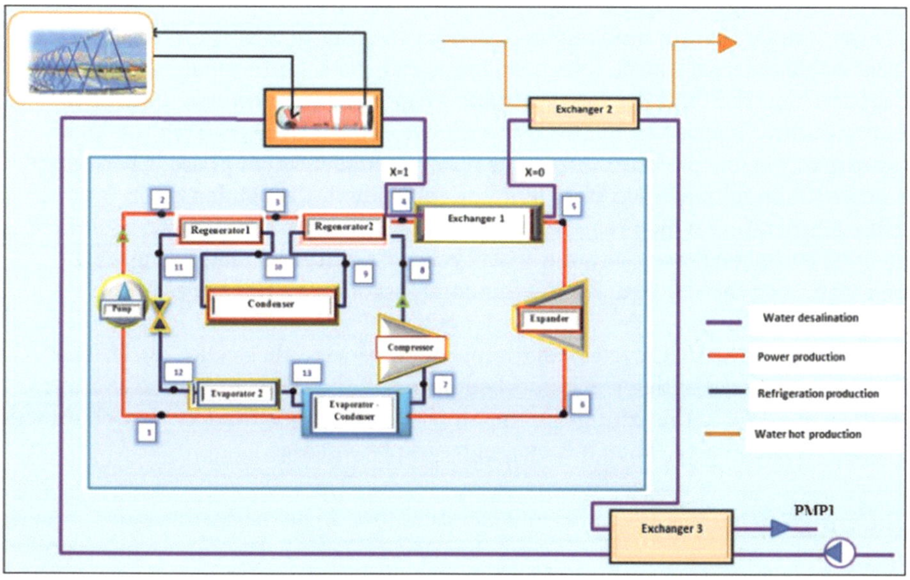

Figure 5.
System of study.

2.1 Desalinated water circuit description

First, the seawater is pumped by a pump PMP1 and preheated by the **exchanger 3**. Then, it will be evaporated at constant pressure in the boiler by a solar collector. After having saturated steam, the latter passed the condensation phase using the **regenerator 2** in order to equate the water at a hot temperature which is equal to that of the evaporation. With integrity of exchanger H2, we started the first phase of cooling the salty water and chaffered sanitary water. In the end, for the desalted water to complete this phase of cooling, also we have to warm the water out of the sea, using an exchanger H3.

2.2 Circuit description of heated domestic water

This is the simplest circuit in our loop. It is enough the sanitary water enters the exchanger H3 to become hot thanks to the quantity of heat provided by the desalted water.

So our system produces electricity due to the mechanical work obtained by the turbine, a refrigeration quantity by the evaporator 1, and de-watered water obtained using two serial transformations (evaporation, condensation) and produces hot water by the exploitation of the hot quantity from the de-watered water.

3. The different configurations developed for ORC-VCC combination

3.1 Configuration A

The cycle A is the basic configuration. We will combine the two ORC and VCC cycles without any recovery for cogeneration. As shown in **Figure 6**, the only combination is made at the heat exchanger H1 which serves as the condenser of the ORC cycle fluid. This configuration allows having cogeneration with positive or negative cold according to our needs.

The operating principle is described as follows:

First, the ORC cycle fluid enters the boiler in order to heat it up to 100°C by a renewable external source (biomass, industrial and solar thermal discharge, etc.). It suffices that the fluid reaches a saturated vapor phase; it enters a turbine to generate a mechanical work. This phase allows the fluid to pass from the high pressure to the low pressure. After this phase, a condensation phase is necessary to make the fluid in a liquid state. For our application, the condensation is done at low temperature which requires a cold external source. For this, we combined the ORC cycle condenser with the VCC cycle evaporator by integrating a H1 exchanger. For this configuration, after condensation, the fluid goes to the pumping phase.

In addition, the VCC cycle operation is the inverse of those ORC cycle. The VCC fluid is compressed with a mechanical compressor and then condensed at a temperature of 30°C. In this configuration, after this phase, the fluid is released directly by an expansion valve. Then it is evaporated in two phases.

3.2 Configuration B

For cycle B, we kept the same basic architecture as in cycle A, except that we will integrate an H2 exchanger. This exchanger is mounted just after the pumping phase of the ORC cycle. Seeing that the temperature obtained at the pumping point is

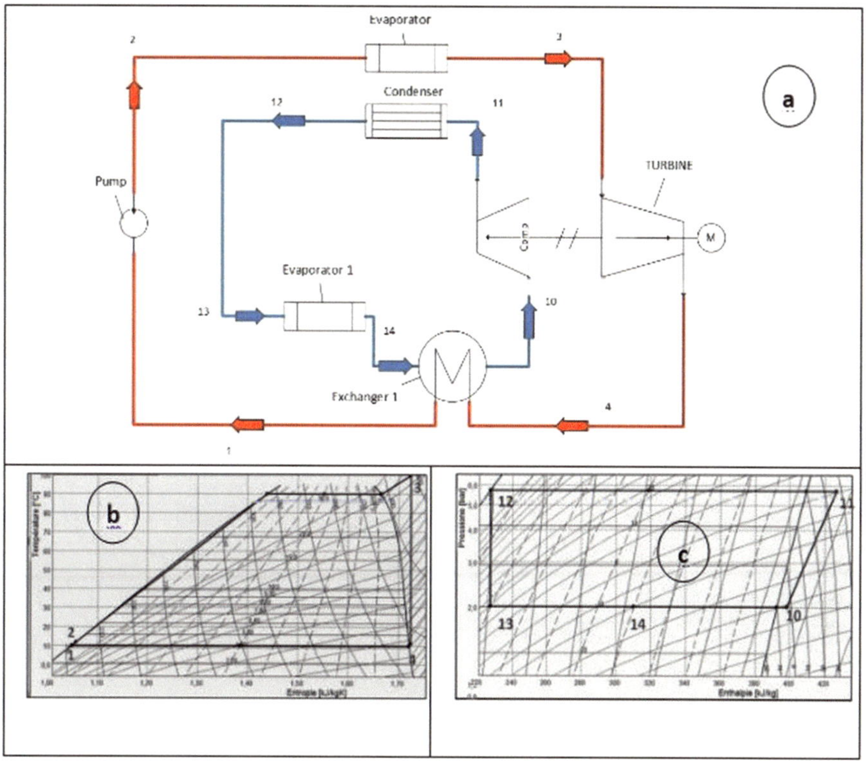

Figure 6.
Schematic and T-S diagrams of the configuration A. (a) Schematic of the configuration A, (b) T-S diagrams for ORC cycle, and (c) P-H diagrams for VCC cycle.

almost the same as the temperature of condensation which varies between −10 and 10°C, the idea is to exploit this temperature to make the sub-cooling of the VCC cycle to improve its performance. Cycle B is shown in **Figure 7**, and it is also developed to make cogeneration with a negative cold.

3.3 Configuration C

As shown in **Figure 8**, cycle C is used also for cogeneration. Unlike the conventional ORC cycle, which is used only for electricity production at the turbine state, cycle C allows the generation of electricity and cold in the ORC cycle. So, the configuration C is used to produce negative cold, positive cold, and electricity.

We will use the heat quantity at low temperature following the pumping step in the ORC cycle in order to produce a positive cold at 18°C for air conditioning. For this reason, we will integrate the H3 exchanger for the heat transfer between the ambient air and the ORC cycle fluid.

4. Mathematical modeling and validation of the model

4.1 Thermodynamic modeling

During this study, we treated the thermodynamic equations as well as the resolution by a calculation program developed by the EES software. Also, this software allows us to realize the different curves and tables presented in the study.

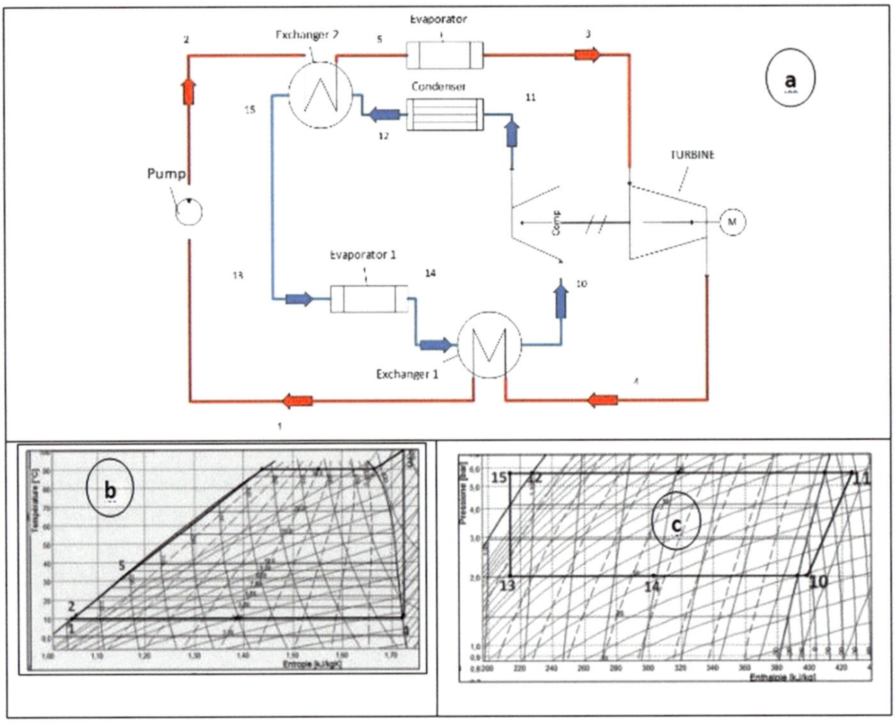

Figure 7.
Schematic and T-S diagrams of the configuration B. (a) Schematic of the configuration B, (b) T-S diagrams for ORC cycle, and (c) P-H diagrams for VCC cycle.

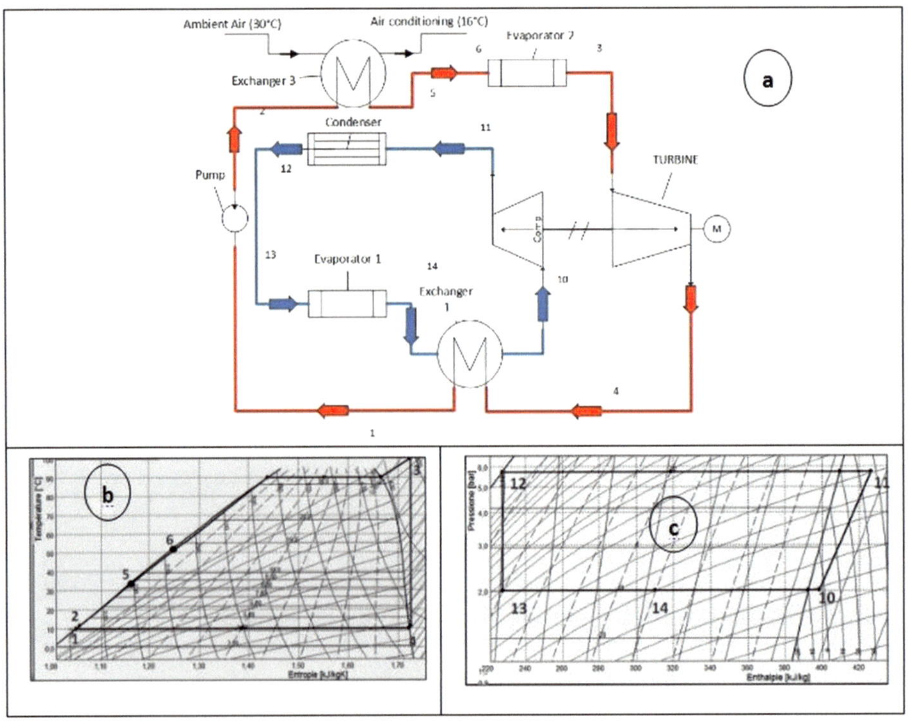

Figure 8.
Schematic and T-S diagrams of the configuration C system. (a) Schematic of the configuration C, (b) T-S diagrams for ORC cycle, and (c) P-H diagrams for VCC cycle.

Table 1 illustrates the different thermodynamic models used throughout the work and the different configurations for cogeneration mode.

4.2 Validation of the model

The approach followed to validate our model is based on two procedures of the developed model. The ORC and the VCC are validated, respectively, in Sections 4.2.1 and 4.2.2.

4.2.1 ORC validation

The model developed for the ORC is tested with the results by Saleh et al. [29], which is the most appropriate configuration to validate the current model using the

ORC	Turbine	$W_T = \dot{m}_1(h_3 - h_{4s}).\eta_{exp}$
	Pump	$W_p = \dot{m}_1 \frac{(h_{2s} - h_1)}{\eta_{pump}}$
	Boiler	(1) $Q_b = \dot{m}_1(h_3 - h_2)$
		(2) $Q_b = \dot{m}_1(h_3 - h_5)$
		(3) $Q_b = \dot{m}_1(h_3 - h_6)$
	Condenser	$Q_{cond} = \dot{m}_1(h_4 - h_1)$
	Thermal efficiency	$\mu_{orc} = \frac{(W_T - W_p)}{Q_b}$
VCC	Compressor	$W_C = \dot{m}_2 \frac{(h_{11s} - h_{10})}{\eta_{comp}}$
	First evaporator	$Q_{ev1} = \dot{m}_2(h_{14} - h_{13})$
	Second evaporator	$Q_{ev2} = \dot{m}_2(h_{10} - h_{14})$
	Overall evaporator	$Q_{ev} = \dot{m}_2(h_{10} - h_{13}) = Q_{ev1} + Q_{ev2}$
	Coefficient of performance	$COP_{VCC} = \frac{(Q_{ev1} + Q_{ev2})}{W_{comp}}$
Overall performance of ORC-VCC	Net work	$W_{net} = W_T - W_p - W_C$
	Overall performance of the system	(1 and 2) $COP_s = \frac{(Q_{ev1} + W_{net})}{Q_b}$
		(3) $COP_s = \frac{Q_{ev1} + W_{net} + Q_{hx3}}{Q_b}$
	Efficacity	(1 and 2) $E = \frac{Q_{ev1}}{W_{net}}$
		(3) $E = \frac{Q_{ev1+} Q_{hx3}}{W_{net}}$
	Exchangers for cogeneration	Exchanger 1 $Q_{hx1} = Q_{ev2} = Q_{cond}$
		Exchanger 2 $Q_{hx2} = \dot{m}_1(h_5 - h_2) = \dot{m}_2(h_{12} - h_{15})$
		Exchanger 3 $Q_{hx2} = \dot{m}_1(h_5 - h_2) = \dot{m}_2(h_{16} - h_{17})$
	Exchangers for tri-generation	$Q_{exh3} = \dot{m}_1(h_5 - h_4) = \dot{m}_3(h_{18} - h_{17})$ $Q_{exh4} = \dot{m}_3(h_{19} - h_{18}) = \dot{m}_4(h_{22} - h_{21})$ $Q_{exh5} = \dot{m}_3(h_{20} - h_{19}) = \dot{m}_3(h_{15} - h_{14})$
Mass ratio		$R_1 = m_3/m_4$ $R_2 = m_1/m_4$ $R_3 = m_2/m_4$ $R_4 = m_3/m_4$

Table 1.
Thermodynamic modeling of different configurations ((1) cycle A; (2) cycle B, and (3) cycle C).

Fluid		T4	P_{min}	P_{max}	m_1	η_{orc}
R600	Ref. [30]	48.43	2.85	15.28	17.746	12.58
	Present model	47.83	2.89	15.52	17.58	12.43
	Error	1.23	1.38	1.54	0.93	1.19
R600a	Ref. [30]	45.33	4.038	19.98	2.423	12.12
	Present model	44.61	4.121	19.79	2.371	11.96
	Error	1.58	2.01	0.95	2.14	1.32
R245fa	Ref. [30]	50.7	1.801	12.67	33.424	12.52
	Present model	49.64	1.765	12.81	34.101	12.44
	Error	2.09	1.99	1.09	1.98	0.63

Table 2.
Validation results for ORC cycle.

similar working applied fluid. The comparative results are illustrated in **Table 2**. These results show a small deviation of 2.09% concerning the thermal efficiency. It is worthy to notice that certain changes in the developed model are made for an appropriate comparison. Specifically, the condensation temperature was 40°C and the isentropic efficiency at 85%.

4.2.2 VCC validation

In this section, the operation of the VCC cycle is enabled. Nasir and Kim [31] are selected for the validation. Some changes in the model are made to have an appropriate comparison against the literature. Indeed, the temperature of the condenser is set to 30°C. **Table 3** includes the validation results along with the COP for cooling. We selected three fluids for validation, which are R245fa, R123, and R134a. **Table 3** shows the margin of error between Ref. [31] and our model. The error results for R245fa, R123, and R134a are, respectively, 0.6, 0.44, and 0.92%. These margins are acceptable to give their low values.

Fluid		COPvcc
R245fa	Ref. [32]	6.60
	Present model	6.56
	Error	0.60
R123	Ref. [32]	6.70
	Present model	6.67
	Error	0.44
R134a	Ref. [32]	6.45
	Present model	6.51
	Error	0.92

Table 3.
Validation results for VCC cycle.

5. Selection of the working fluid

The choice of the working fluid for an ORC or VCC cycle is an important criterion to improve the cycle performances. Generally, there are three families of organic fluids. **Figure 9** shows these three classes on a T-S diagram. The distinction between these different types essentially depends on the slope between the saturation temperature and the isentropic variation ($\Delta T/\Delta s$). If a negative slope is said, the fluid is wet, such as H_2O, NH_3, and R134a. For a positive slope, we speak of a dry fluid such as benzene and pentane. In cases where the slope is infinite, it is said that this fluid is isentropic like R600 and R600a.

For the ORC cycle is to have fluid hot admits a weak latent heat in the evaporator, in order to minimize the quantity received by the boiler. Thus, a low latent heat in the condenser minimizes the amount of cold delivered by the VCC cycle. In addition, we are looking for a fluid with a positive slope to avoid vapor having less than 0.95 of steam rate.

We guarantee the elimination of the oxidation effect in the turbine especially that we will make it lower concerning the condensation temperature to −10°C. Based on these criteria and conditions mentioned above, it is necessary to choose a dry or isentropic ORC cycle fluid. We choose the n-hexane; the chemical formula is C_6H_{14}. The thermophysical characteristics of this fluid are presented in **Table 4**.

The R600 is selected as a working fluid for the VCC cycle. It is a hydrocarbon of formula C_4H_{10} crude which is found in the gas status under normal conditions of temperature and pressure. The physical characteristics of this fluid are presented in **Table 4**. Furthermore, our choice is toward the use of n-hexane for the ORC cycle. This choice is essentially due to the steam rate which is equal to 1 even when the condensation temperature is lowered to a low degree. This allows us to have a margin of confidence and turbine safety (avoid the effect of oxidation). During our study, we chose the R600 as a working fluid for the VCC cycle. This fluid is characterized by its robustness in the market, so it is used in recent years in several researches. In addition, we find that the environmental damage is minimal.

6. System settings and boundary conditions

To reassure the efficiency and rentability of the system, it is necessary that we set some parameters and define their limits. For example, the network and

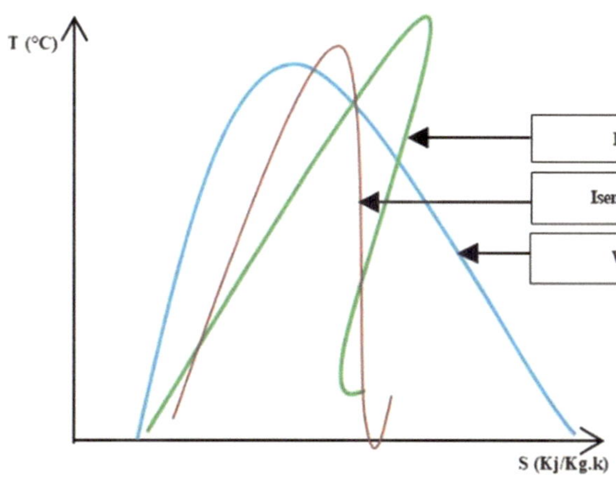

Figure 9.
The three classes on a T-S diagram.

Fluid	Critical temperature (°C)	Critical pressure (bar)	MW (kg/kmol)
Ammonia	132.3	113.3	17.03
R600a	134.7	36.4	58.12
R134a	101	40.59	102
R500	105.5	44.55	99.31
R236fa	124.9	32	152
Propane	96.68	42.47	44.1
R245fa	154	36.51	134
Acetone	235	47	58.08
n-Hexane	234.7	30.58	86.17
R600	152	37.96	58.12
R123	183.7	36.68	152.9

Table 4.
Physical and chemical property of work fluids.

refrigeration capacity must be always positive. Also, to guarantee the safety of the turbines, it is necessary that the vapors' quantity must be more than 95%.

The boundary conditions are shown in **Table 5**.

7. Results analysis and discussion

The main purpose of this study is to analyze the performance of a new system that combines the steam compression cycle and the Rankine cycle for tri-generation (electricity, cold, hot).

In the previous section, we presented the different designs of the system. This system has an energy autonomy. It needs only the solar temperature "T_h." For that reason, we will focus our work on the impact of solar temperature.

First of all, we started with the mass flow analysis of each circuit in order to have the mass potential of our mini power plant.

Figure 10 resumes the evaluation of the different values of the flow rate for each circuit. The different ratios of the mass flow rates are R_1, R_2, R_3, and R_4. The shape of different curves is of positive exponential form. It can be seen that the variation between the four curves is constant in function of T_h. The three ratios R_1, R_3, and R_4 represent small variations of the ratios of the flow rates as a function of T_h between them. Consequently, the geometries of different constituent bodies are coherent in terms of dimensions. In contrast, the ratio R_2 is a large margin of variation compared to the other ratios.

Wnet	>0
Qevnet	>0
X4	>0.95
$Tair_{in}$	30
$Tair_{out}$	18
T_6	30

Table 5.
Boundary conditions.

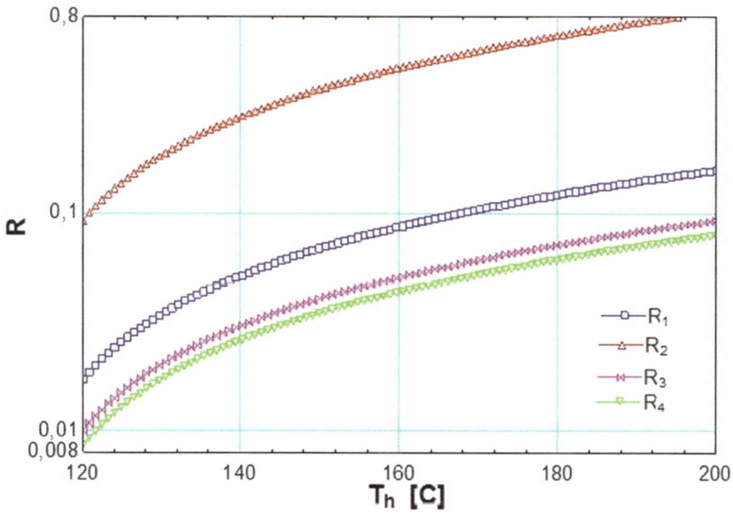

Figure 10.
The evolution of different throughput ratios as a function of T_h.

This variation in flow rates does not depend only on the temperature of the solar collector T_h, but it also depends on the temperature of the boiler T_g. For this, we have made surfaces of each flow ratio with two temperatures T_h and T_g as shown in **Figure 11**.

In addition, it can be noticed that the net quantity of the hot water is delivered by the system.

Figure 12 illustrates the evolution of different mass flow rates as a function of the heat delivered Q_b by the boiler. The four flows are varied proportionally with Q_b. It is

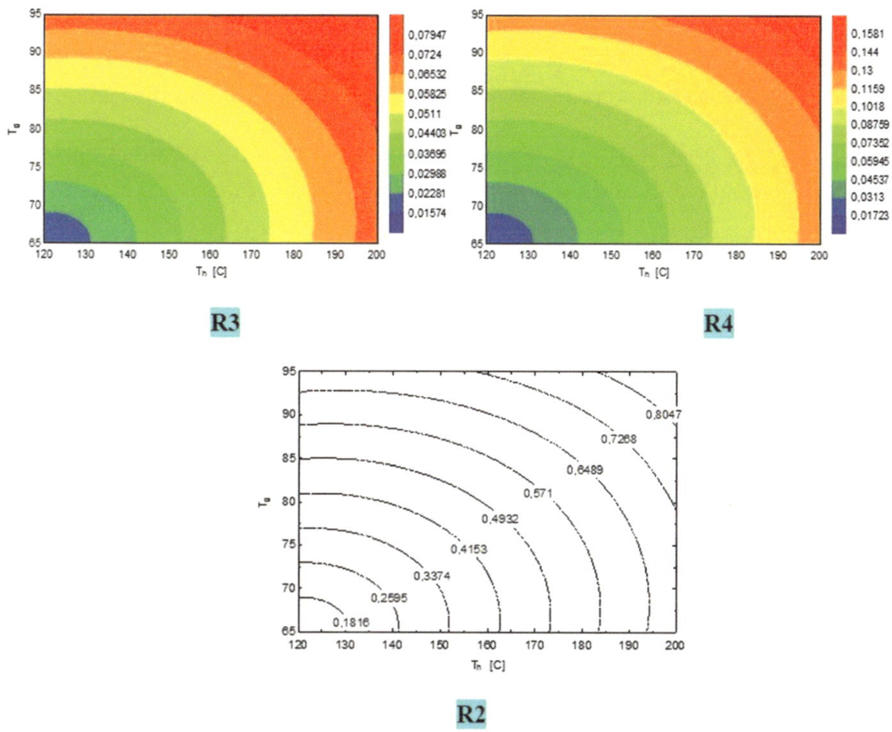

Figure 11.
Surfaces of flow reports as a function of each two temperatures T_h and T_g.

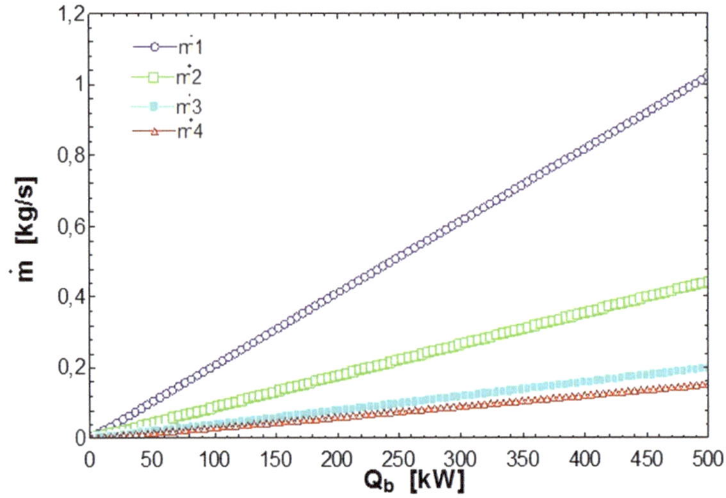

Figure 12.
The evolution of flows according to high-temperature heat source.

noted that the mass flow has a large positive slope with respect to the other mass flow rates. This allows us to interpret that the geometry of the ORC cycle is very important in relation to the different cycles. Also, this cycle promotes significant power.

It is constant that the lowest mass flow rate corresponds to the mass flow rate that does not exceed 0.18 kg/s, which means that the desalinated water is installed at a low power.

In energy potential term provided by our installation, **Figures 13** and **14** indicate the net work and the amount of cold produced as a function of the hot source Q_b. It is observed that the two quantities W_{net} and Q_{ev} are proportional to Q_b. It is possible to obtain a net work of maximum value equal to 14 kW and a maximum amount of cold equal to 75 kW.

Figure 15 shows the evolution of the ORC thermal efficiency as a function of the generator temperature T_g and the vaporization temperature of the cold T_{ev}.

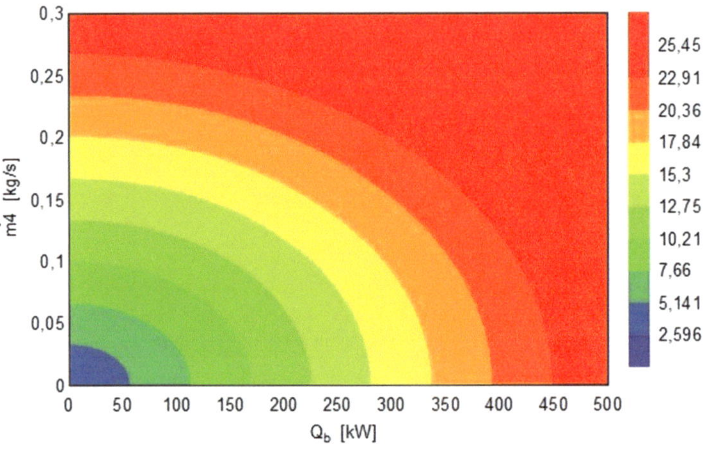

Figure 13.
The evolution of flows m_4 according to high-temperature heat source.

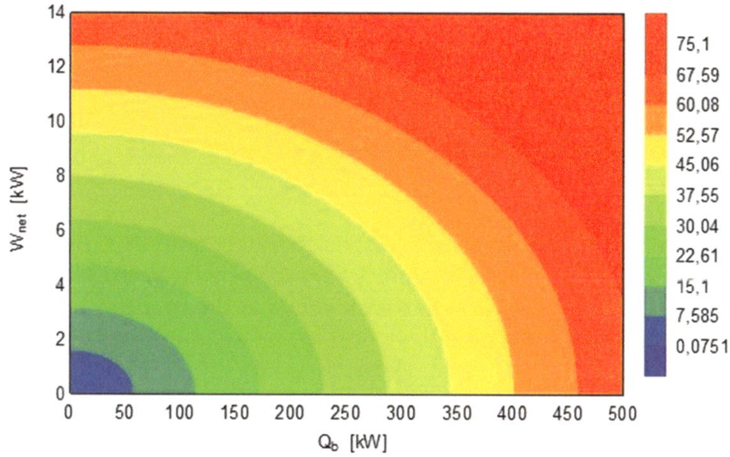

Figure 14.
Net work variation and cooling capacity according to Q_b.

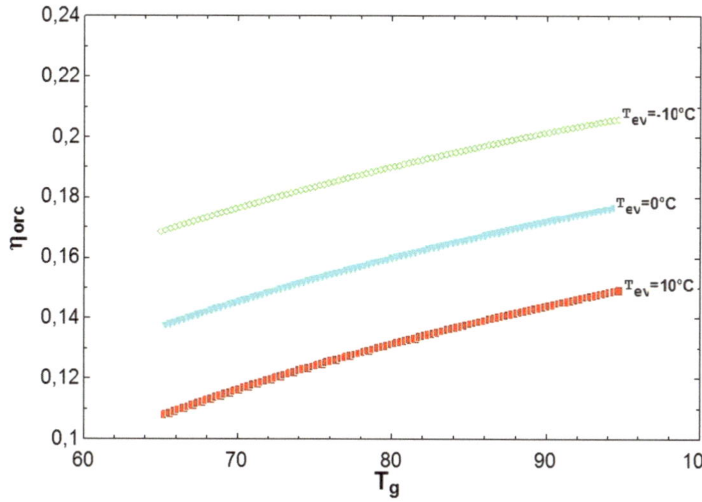

Figure 15.
Variation of ORC efficiency as a function of T_g and T_{ev} temperature.

It is observed that the ORC efficiency is proportional to T_g and inversely proportional to T_{ev}; this is justified by the first principle of thermodynamics. A better efficiency noted is 0.21 for $T_{ev} = -10°C$ and $T_g = 95°C$.

7.1 Technical-economic analysis and investment costs

The investment costs can be estimated from specialized works [30, 32–36] where they are generally presented in the form of charts or tables. Often these abacuses represent the cost taking into account the influence of parameters such as pressure, temperature, material or manufacturing, assembly, transport, etc.

We have undertaken to gather technical and economic data of the components used in the VCC and ORC cycles (compressors, condensers, evaporators) in order to develop technical-economic and then exergo-economic con-elations. This task is not easy since often the data is discrete and the interpolations or extrapolations are not conclusive due to the nonlinearity of the cost according to the parameters used by the manufacturer.

7.1.1 Compressor cost

For compressors, the price depends on the type of compressor, the power, and the volume swept, while for motor compressors the price is given according to the type, the power, and flow of the heat transfer fluid at the condenser.

Compressor cost = f (type, power, volume swept).

Motor compressor cost = f (type, power, flow rate of the coolant at the condenser).

So for each model corresponding to a certain type of compressor, a function of the following form is established:

Compressor cost = a. $(\text{Power})^b$. $(\text{Volume swept})^c$

Determine the coefficients a, b, and c.

We are able to express by an analytical approach the price of a certain type of compressor, knowing the technical characteristics.

Generally correlations have been developed to determine the investment costs of each organ.

For the compressor, the correlations used are [37]:

$$ci\_{CP} = 9.27 . 10^{-8} \; cyl^{(-4.85)} \; \dot{Q}_{EV}^{33} \; \dot{W}^{2}$$

with

$10.3 < cyl < 30.6$ [cm^3]

$835 < Q_{EV} < 2650$ [W]

$475 < W < 1225$ [W]

7.1.2 Evaporator cost

For evaporators the price depends on the type of evaporator, the power, the exchange surface, the flow of the heat transfer fluid at the evaporator, the number of fans, and the pitch of the fins.

Evaporator price = function (type, power, surface, flow, no. of fans, no fins).

For the evaporator, the correlations used are [37]:

$$ci\_{EV} = 2.94 \; \dot{Q}_{EV}^{0.44} \; D_{EV}^{0.42}$$

with

$850 < Q_{EV} < 5500$ [W]

$685 < D_{EV} < 3325$ [m^3/h]

7.1.3 Condenser cost

The price of a condenser is given according to the type of condenser, the power, the exchange surface, the flow of heat transfer fluid, and the number of fans.

Price = function (type, power, surface area, heat transfer fluid flow, no. of fans).

For the condenser, the correlations used are [37]:

$$ci\_{CD} = 2.4 \cdot 10^5 \, \dot{Q}_{CD}^{-0.53} \, D_{CD}^{1.96}$$

with
$3500 < Q_{CD} < 20{,}000$ [W]
$0.6 < D_{CD} < 2$ [m^3/h]

8. Conclusion

The energy performance of power and refrigeration cogeneration and tri-generation through an organic Rankine cycle (ORC) with a vapor compression cycle (VCC) by a new combination systematic is examined. We can use a low-temperature energy source. Two cases of refrigeration and cogeneration are analyzed, including cases of cogeneration (−10, 10°C) and congelation (0, −17°C).

The effects of the system parameters include the condensation and vaporization temperatures for ORC and VCC, and the efficiency E on performance such as thermal efficiency, specific refrigeration, and net work output and global system performance are investigated.

According to the analysis and the investigation carried out during this study, the main interpretations retained are:

- The results show that operating parameters have a significant effect on performance. This effect differs from one use case to another (positive or negative refrigeration) and according to the installed configuration (cycles A, B, and C).

- The three configurations developed which were based on the integration of recovery exchangers noted improvements in overall performance. These improvements also differ from one cycle to another, which makes it possible to say that the spot of integration of the exchangers is an effect on the performances.

- The results show that for cogeneration with negative cold, among the three configurations that we have developed, cycle B is preferable in which it has a better energy performance.

Nomenclature

COP_{vcc}	coefficient of performance for the vapor compression cycle
COP_s	coefficient of performance for the overall system
Ci_{cp}	investment cost of the compressor
Ci_{EV}	investment cost of the evaporator
Cyl	piston compressor displacement
Ci_{CD}	investment cost of the condenser
D_{CD}	heat transfer fluid flow rate at condenser
D_{EV}	heat transfer fluid flow rate at evaporator
H1	exchanger 1
H2	exchanger 2
H3	exchanger 3
+ H2. H1	cycle with exchangers 1 and 2

H	enthalpy (kJ/kg)
MW	molar mass
m_1	mass flow rate for ORC (kg/s)
m_2	mass flow rate for VCC cycle (kg/s)
m_3	mass flow rate for heated water
m_4	mass flow rate for seawater desalination
ORC	organic Rankine cycle
Pcrit	critical pressure
P_{sat}	saturated pressure
R_m	mass flow ratio
R_{pp}	pressure ratio for pump
R_{pc}	pressure ratio for compressor
T	temperature (°C)
T_{ev}	vaporization temperature for VCC cycle (°C)
T_{cit}	critical temperature (°C)
T_h	temperature in the panel solar (°C)
T_g	temperature in the boiler for ORC cycle
T_{cond}	condensation temperature for organic Rankine cycle
T_{sh}	overheating temperature for organic Rankine cycle
Q_b	boiler heat input (kW)
Q_{h2}	heat input for the exchanger 2 (kW)
Q_{ev1}	the power of the evaporator 1 (kW)
Q_{ev2}	the power of the evaporator 2 (kW)
Q_{ev}	the overall power evaporated by the VCC cycle (kW)
Q_{EV}	heat flow exchanged at the evaporator
Q_{CD}	heat flow exchanged at the condenser
VCC	vapor compression cycle
W_{com}	working fluid pump power consumption (kW)
W_{exp}	expander work output (kW)
W_{net}	net work output for overall system (kW)
W_{pump}	working fluid pump power consumption (kW)
W	compressor power
W_T	mechanical work of the turbine (KW)
Wc	mechanical work of the compressor (kW)
X	title vapor
η_{is1}	compressor isentropic efficiency
η_{is2}	expander isentropic efficiency
η_{pump}	working fluid pump isentropic efficiency
ΔT_{Pinch}	pinch temperature (°C)

Index

1	pump inlet
2	pump outlet
3	boiler output and expander inlet
4	expander outlet
10	compressor inlet
11	compressor outlet
11	condenser inlet
12	condenser outlet
13	evaporator 1 inlet
14	evaporator 1 outlet

Author details

Noureddine Toujani*, Nahla Bouaziz and Lakder Kairouani
Energetic and Environmental Research Unity, National Engineering School of Tunis, Tunis El Manar University, Tunis, Tunisia

*Address all correspondence to: toujeninoureddine@gmail.com

References

[1] International Energy Agency (IEA). International Energy Outlook-Highlights. Washington DC, USA: IEA; 2010

[2] Mohanty S. Forecasting of solar energy with application for a growing economy like India: Survey and implication. Renewable and Sustainable Energy Reviews. 2017;**78**:539-553

[3] Nematollahi O. A feasibility study of solar energy in South Korea. Renewable and Sustainable Energy Reviews. 2017; **77**:566-579

[4] Ozoegwu CG. The status of solar energy integration and policy in Nigeria. Renewable and Sustainable Energy Reviews. 2017;**70**:457-471

[5] Sadiq AA, Dada Joseph O, Khalil AI. Current status and future prospects of renewable energy in Nigeria. Renewable and Sustainable Energy Reviews. 2015; **48**:336-346

[6] Herche W. Solar energy strategies in the U.S. utility market. Renewable and Sustainable Energy Reviews. 2017;**77**: 590-595

[7] Communication from the Commission to the European Parliament and the Council: "Energy Efficiency 610 and its contribution to energy security and the 2030 Framework for climate and energy policy"-Brussels, 611 23.7.2014 COM (2014) 520 final

[8] Sarbu I, Sebarchievici C. General review of solar-powered closed sorption refrigeration systems. Energy Conversion and Management. 2015;**105**: 403-422

[9] Gingerich DB, Mauter MS. Quantity, quality, and availability of waste heat from United States thermal power generation. Environmental Science and Technology. 2015;**49**:8297-8306

[10] Chen CL, Li PY, Le SNT. Organic Rankine cycle for waste heat recovery in a refinery. Industrial & Engineering Chemistry Research. 2016;**55**:3262-3275

[11] Sansaniwal SK, Sharma V. Energy and exergy analyses of various typical solar energy applications: A comprehensive review. Renewable and Sustainable Energy Reviews. 2018;**82**:1576-1601

[12] Bolaji B. Exergetic analysis of solar drying systems. Natural Resources. 2011;**2**(2):92-97

[13] Fudholi A, Sopian KB, Othman MY, Ruslan MH. Energy and exergy analyses of solar drying system of red seaweed. Energy and Buildings. 2014;**68**(Part A): 121-129

[14] Gunhan T, Ekren O, Demir V, Sahin AS. Experimental exergetic performance evaluation of a novel solar assisted LiCl–H_2O absorption cooling system. Energy and Buildings. 2014;**68** (Part A):138-146

[15] Siddiqui FR, El-Shaarawi MAI, SAM S. Exergo-economic analysis of a solar driven hybrid storage absorption refrigeration cycle. Energy Conversion and Management. 2014;**80**:165-172

[16] Bouaziz N, Lounissi D. Energy and exergy investigation of a novel double effect hybrid absorption refrigeration system for solar cooling. International Journal of Hydrogen Energy. 2015; **40**(39):13849-13856

[17] Gang P, Guiqiang L, Xi Z, Jie J, Yuehong S. Experimental study and exergetic analysis of a CPC-type solar water heater system using higher-temperature circulation in winter. Solar Energy. 2012;**86**(5):1280-1286

[18] Shukla SK, Gupta SK. Performance evaluation of concentrating solar cooker

under Indian climatic conditions. In: Proceedings of the Second International Conference on Energy Sustainability; Jacksonville, Florida, USA; 10–14 Aug, 2008

[19] Naik PS, Palatel A. Energy and exergy analysis of a plane reflector integrated photovoltaic-thermal water heating system. ISRN Renewable Energy. 2014; **2014**:9. Article ID 180618. DOI: 10.1155/2014/180618

[20] Wu J, Zhu D, Hua W, Zhu Y. Exergetic analysis of a solar thermal power plant. Advances in Materials Research. 2013;**724–725**:156-162

[21] Ehtiwesh IAS, Coelho MC, Sousa ACM. Exergetic and environmental life cycle assessment analysis of concentrated solar power plants. Renewable and Sustainable Energy Reviews. 2016;**56**:145-155

[22] Cau G, Cocco D. Comparison of medium size concentrating solar power plants based on parabolic through and linear Fresnel collectors. Energy Procedia. 2014;**45**:101-110

[23] Chang K-H, Lin G. Optimal design of hybrid renewable energy systems using simulation optimization. Simulation Modelling Practice and Theory. 2015;**52**:40-51

[24] Abedi S, Alimardani A, Gharehpetian G, Riahy G, Hosseinian S. A comprehensive method for optimal power management and design of hybrid RES-based autonomous energy systems. Renewable and Sustainable Energy Reviews. 2012;**16**:1577-1587

[25] Olatomiwa L, Mekhilef S, Ismail M, Moghavvemi M. Energy management strategies in hybrid renewable energy systems: A review. Renewable and Sustainable Energy Reviews. 2016;**62**: 821-835

[26] Torreglosa JP, García-Triviño P, Fernández-Ramirez LM, Jurado F. Control based on techno-economic optimization of renewable hybrid energy system for stand-alone applications. Expert Systems with Applications. 2016;**51**:59-75

[27] Dash V, Bajpai P. Power management control strategy for a stand-alone solar photovoltaic-fuel cell–battery hybrid system. Sustainable Energy Technologies and Assessments. 2015;**9**:68-80

[28] Zhou T, François B. Energy management and power control of a hybrid active wind generator for distributed power generation and grid integration. IEEE Transactions on Industrial Electronics. 2011;**58**:95-104

[29] Saleh B, Koglbauer G, Wendland M, Fischer J. Working fluids for low-temperature organic Rankine cycles. Energy. 2007;**32**:1210-1221

[30] Chauvel A et al. Manuel d'Evaluation Economique des Procédés. Paris: Edition Technip; 1976

[31] Nasir MY, Kim KC. Working fluids selection and parametric optimization of an organic rankine cycle coupled vapor compression cycle (ORC-VCC) for air conditioning using low grade heat. Energy and Buildings. 2016;**129**:378-395

[32] Chatelain C, Ducrocq JC, Mignard B, Coeytaux M. Optimisation technico-economique des processus énergétiques. Technique de l'Ingénieur, Génie Énergétique, B. 1996;**1282**:1-24

[33] Peters MS, Merhaus KD. Plant Design and Economics for Chemical Engineers. 4th ed. New York: McGraw Hill; 1991

[34] Garett DE. Chemical Engineering Economics. New York: Van Nostrand Reinhold; 1989

[35] Baasel WD. Preliminary Chemical Engineering Plant Design. 2nd ed. New York: Van Nostrand Reinhold; 1990

[36] Guthrie KM. Process Plant Estimating, Evaluating and Control. Solana Beach, CA, USA: Craftsman; 1974

[37] Lavinia GROSU. Contribution a l'optimisation thermodynamique et économique des machines a cycle inverse a deux et trois reservoirs de chaleur. Available from: http://www.cfcopies.com/V2/leg/leg_droi.php

Chapter 2

Endomembrane Trafficking in Plants

Birsen Cevher-Keskin

Abstract

The functional organization of eukaryotic cells requires the exchange of proteins, lipids, and polysaccharides between membrane compartments through transport intermediates. Small GTPases largely control membrane traffic, which is essential for the survival of all eukaryotes. Transport from one compartment of this pathway to another is mediated by vesicular carriers, which are formed by the controlled assembly of coat protein complexes (COPs) on donor organelles. The activation of small GTPases is essential for vesicle formation from a donor membrane. In eukaryotic cells, small GTP-binding proteins comprise the largest family of signaling proteins. The ADP-ribosylation factor 1 (ARF1) and secretion-associated RAS superfamily 1 (SAR1) GTP-binding proteins are involved in the formation and budding of vesicles throughout plant endomembrane systems. ARF1 has been shown to play a critical role in coat protein complex I (COPI)-mediated retrograde trafficking in eukaryotic systems, whereas SAR1 GTPases are involved in intracellular coat protein complex II (COPII)-mediated protein trafficking from the endoplasmic reticulum (ER) to the Golgi apparatus. The dysfunction of the endomembrane system can affect signal transduction, plant development, and defense. This chapter offers a summary of membrane trafficking system with an emphasis on the role of GTPases especially ARF1, SAR1, and RAB, their regulatory proteins, and interaction with endomembrane compartments. The vacuolar and endocytic trafficking are presented to enhance our understanding of plant development and immunity in plants.

Keywords: GTPases, ARF1 (ADP-ribosylation factor 1), SAR1 (secretion-associated RAS superfamily 1), COPI (coat protein complex I), COPII (coat protein complex II), membrane traffic, clathrin

1. Introduction

Endomembrane trafficking plays a crucial role for maintaining fundamental cellular functions (signal transduction, cellular homeostasis, etc.) and in response to environmental stimuli. The membrane trafficking pathways start from the endoplasmic reticulum (ER) then go through the Golgi apparatus to different destinations including vacuoles/lysosomes, endosomes, and the plasma membrane (PM) [1].

In plant cells the membrane trafficking system comprises three major trafficking pathways: the biosynthetic secretory pathway, the endocytic pathway, and the vacuolar transport pathway. (i) *The biosynthetic secretory pathway* transports newly synthesized proteins from the endoplasmic reticulum to the plasma membrane and/or the extracellular space. (ii) *The endocytic pathway* functions in the recycling of PM-localized and extracellular factors between the PM and the endosomal compartments. (iii) *The vacuolar transport pathway* drives the transportation of newly synthesized protein to the vacuole (**Figure 1**) [3].

Each trafficking pathway is mediated by the following steps: (i) budding of the transport vesicle from the donor membrane, which is mediated by ARF/SAR1 GTPase (and coat proteins in many cases); (ii) transport and targeting of the transport vesicle; (iii) tethering of the vesicle by tethering proteins under the regulation of RAB GTPase and fusion of the transport vesicle to the target membrane mediated by soluble N-ethylmaleimide-sensitive-factor attachment protein receptors (SNARE); and (iv) recycling of the transport machinery components to the donor membrane (**Figure 2**).

In eukaryotic cells, small GTP-binding proteins involve the largest family of signaling proteins. The activation of small GTP-binding proteins (GTPases) is essential for vesicle formation from a donor membrane. Four main subfamilies have been identified in plants: (i) ADP-ribosylation factor (*ARF*)/secretion-associated RAS superfamily (*SAR*), (ii) *RAB*, (iii) Rho-like proteins in plants (*ROP*), and (iv) *RAN* [5–7]. Over the evolution of eukaryotic organisms, the conservation of GTPases explains their significance in cellular signaling processes [7–9]. Small GTPases serve as molecular switches that transduce signals by exchanging between the GTP- and GDP-bound conditions. Guanine nucleotide exchange factors (GEFs), GDP dissociation inhibitors (GDIs), and GTPase-activating proteins are regulators of small GTP-binding proteins.

GEFs activate small GTPases, which in turn interact with specific effectors to stimulate downstream pathways. GAPs trigger the intrinsic GTPase activity, thereby accelerating the inactivation of the GTPases' regulatory activity. GEFs convert the GDP-bound inactive form of the GTPases to the GTP-bound active form by stimulating the dissociation of GDP from the GDP-bound form. In the "active" state, the GTP-bound GTPases interact with various downstream effector proteins that execute diverse cellular functions. GTPases are inactivated through either the intrinsic capability of the GTPase to hydrolyze GTP to GDP + Pi or an interaction with another protein group, the GTPase-activating proteins (GAPs). These proteins

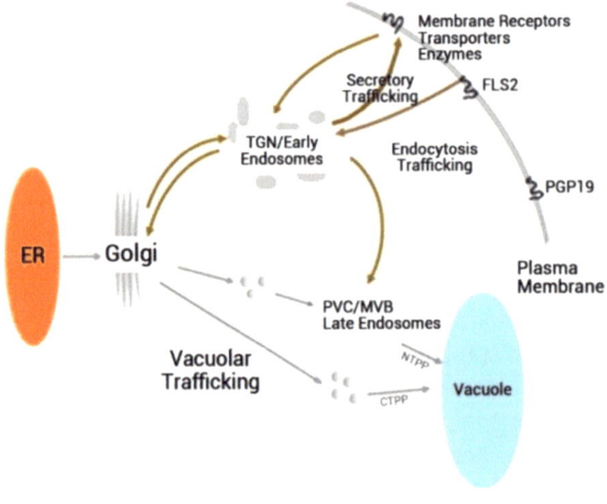

Figure 1.
The membrane trafficking pathways are grouped into three major categories: (i) the biosynthetic secretory pathway, (ii) the endocytic pathway, and (iii) the vacuolar transport pathway (modified from Inada and Ueda [2]).

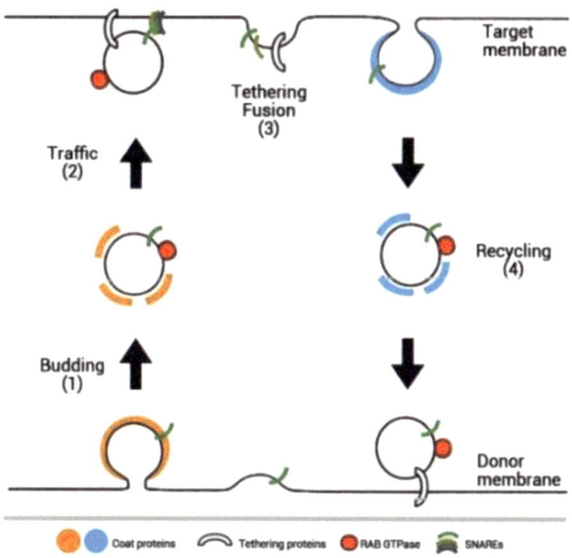

Figure 2.
The general machinery of membrane trafficking. Each trafficking pathway is mediated by the following steps: (i) budding of the transport vesicle from the donor membrane, which is mediated by ARF/SAR1 GTPase (and coat proteins in many cases); (ii) transport and targeting of the transport vesicle; (iii) tethering of the vesicle by tethering proteins under the regulation of RAB GTPase and fusion of the transport vesicle to the target membrane mediated by SNARE proteins; and (iv) recycling of the transport machinery components to the donor membrane [2, 4].

catalyze the hydrolytic activity of GTPases, which then return to the inactive state GDP-bound state [6]. The improvement of fluorescent protein-labeled GTPases and cargo molecules has enhanced the assignment of subcellular locations for these proteins within the endomembrane system.

The traffic between organelles is bi-directional. (i) Starting at the ER and leading toward the destination organelles (the forward trafficking) is called anterograde transport, and (ii) the reverse pathway is called retrograde transport.

2. Membrane trafficking pathways

2.1 The biosynthetic secretory pathway

2.1.1 Anterograde transport (forward trafficking)

2.1.1.1 ER-to-Golgi protein transport

The conventional trafficking pathway starts at the ER; protein synthesis and modification occurs and undergoes further modification [10]. Proteins leave the ER via COPII carriers to reach the Golgi. ER and Golgi compartments are closely associated with each other to ease the movement of cargo between them [11]. This ER-to-Golgi transport is termed "anterograde transport" and is mediated by COPII proteins, which are highly conserved in eukaryotes [12]. From the ER, synthesized proteins are exported to the cis-Golgi and are transported via Golgi stacks where protein modifications occur. Modified proteins are sorted into the extracellular space or storage and lytic organelles from the Golgi. In plants, proteins can also be sorted from the Golgi into the chloroplast [13].

The accumulation of secretory cargo, deformation of the membrane, and formation of transport vesicles are mediated by COPII. In mammalian cells and in most of the plant cell types, the ER and Golgi are in close proximity, and the COPII cycles on and off the ER with a fast turnover rate [14, 15]. COPII coat proteins are mostly distributed in the cytosol and concentrate at ERES that appear in association with motile Golgi stacks in plant cells. These proteins also accumulate in punctate structures that are not associated with the Golgi (**Figure 2B**).

The recruitment of COPII coat proteins involves SAR1 GTPase and its GDP/GTP exchange factor, SEC12 [16, 17]. The *Arabidopsis* genome encodes five genes for SAR1, seven genes SEC23, three genes for SEC24, two genes SEC13, and two SEC31 isoforms [18].

The assembly of COPII occurs at distinct sites on ribosome-free transitional ER (tER) or ER exit sites (ERESs) [19]. The cytosolic GTPase SAR1 is activated by the ER membrane-associated GEF SEC12, and then SAR1 associates with the ER lipid bilayer membrane, and after the COPII coat composed of the SEC23-24 and SEC13-31 heterodimer complexes is recruited [20–22]. The cargo recruitment complex involving SEC23-24 and SAR1 sorts transport and ER resident proteins [23, 24]. The COPII coat includes four proteins, assembled as an internal receptor/cargo-binding dimer of SEC23 and SEC24 and an outer cage dimer of SEC31 and SEC13. SEC16 is important for ER protein export by recognizing the COPII assembly region at the ERES [25]. The cargo selection is achieved by the SEC23/SEC24-SAR1 complex (pre-budding complex) [26]. This complex recruits SEC13-SEC31, which offer the outer layer of the coat and manage membrane deformation to constitute COPII vesicles. The SEC16 and SED4 are the other additional proteins for the COPII assembly. SEC16 comprises COPII coat component domains and has an important role as a scaffold for coat assembly [27]. SEC16 is a key organizer of ERESs in yeast and mammalian cells [28, 29]. The two encoded from *Arabidopsis* SEC16 genes resemble the human small isoform, and it was shown that they are important for ER export and tER organization in HeLa cells [29].

COPII is also involved in the physical deformation of the ER membrane that drives the COPII carrier formation [30]. SAR1-mediated GTP hydrolysis leads to COPII carrier un-coating and follows the exposure of the carrier membrane to fusion with the Golgi membrane [31].

In the GTP-bound conformation, SAR1 protein binds directly to the lipid bilayer which it does by an N-terminal amphipathic alpha-helix [32]. In the GDP-bound conformation, SAR1 binds membranes with lower affinity [25, 31].

The SED4 is responsible for the rate of ER-to-Golgi transport as an integral membrane protein at the ER membrane [32]. The deletion of SED4 causes to reduce the transportation rate of ER-to-Golgi in *S. cerevisiae* wild-type cells [33]. The SED4 and SEC12 have close homology with the cytoplasmic domain, but no GEF activity has been reported in *S. cerevisiae* [34].

It has been reported that SAR1 reduces the mechanical rigidity of the lipid bilayer membrane to which it binds in yeast [35]. Because of the ability of SAR1, it was suggested that membrane-bound SAR1-GTP decreases the energetic cost for the other COPII coat proteins (Sec13, Sec31, etc.) to generate curvature [35].

In *Arabidopsis*, three SAR1 homologs have been identified (AtSARA1a, AtSARA1b, and AtSARA1c). AtSARA1A and AtSARA1B have a 93% amino acid sequence identity [36]. The AtSARA1a expression level correlated with the secretion activity level from ER membranes. The AtSARA1a mRNA upregulation has been reported to cause the blockage of ER transport to the cis-Golgi compartment [37]. The COPII protein-encoding genes are ubiquitously expressed except SAR1 (At1g09180) and a SEC31 (At1g18830) isoform by microarray analyses [18]. Tissue specificity was observed for the SEC31 isoform At1g18830, while all other genes appear to be ubiquitously expressed in all tissues and developmental stages [18].

SAR1 accumulation was observed to concentrate predominantly in crude ER fractions of *Pisum sativum* L. seedlings [38]. The COPII protein coat recruitment by SAR1p has been intensively studied [39]. In human development and disease, the SEC24 and SAR1 isoforms have specificity for the trafficking of selective cargo [40, 41]. In plant cells, specific amino acid sequences in the primary proteins affect the selective export of membrane cargo [42]. Diacidic sequence induces the accumulation of SEC24A to ERESs. The interaction occurs between the K channel KAT1, which contains the specific amino acid sequence in the cytosolic tail, and SEC24A [43, 44].

Different export signals in SEC24 proteins might lead to selective accumulation of cargo in COPII carriers in mammalian and plant cells. It was suggested that more efficient intracellular trafficking is likely achieved by cargo specialization of COPII isoforms in multicellular organisms [45]. **Figure 3** shows the basic diagram of the retrograde and anterograde transport in plant cells [47].

The COPII machinery has significant importance for ER-to-Golgi transport in the early secretory pathway in plants [48]. The retention of secretory cargo molecules or membrane proteins that cycle between the ER and Golgi apparatus leads to blockage of the ER export [14, 43, 48]. COPII machinery is involved in biotic and abiotic stress responses in plants. It has been shown that functional SEC24A is essential for systemic turnip mosaic virus movement by interaction in a signal specific manner with the N-terminal domain 6-kDa viral protein 6 K [49]. In high-temperature conditions, overexpression of SEC31A has been reported in the IRE1 mutant which leads to improvement of the male sterility phenotype in *Arabidopsis* [50].

2.1.2 Retrograde transport (reverse trafficking)

2.1.2.1 Retrograde transport and COPI

Of the 12 ARF isoforms, ARF1 is targeted to the Golgi and post-Golgi structures in plant cells. *Arabidopsis* ARF1 has been shown to be involved in different trafficking pathways including ER-Golgi traffic, vacuolar trafficking, and endocytosis and/or recycling [51, 52]. Arfs are divided into three classes and express six isoforms, namely, Arf1 to Arf 6 (with Arf2 being absent in human) in the mammalian system.

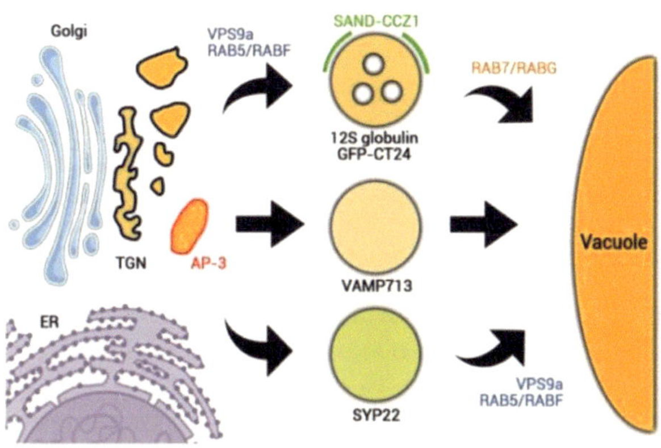

Figure 3.
Model of membrane trafficking to the vacuole in plant cells. The vacuolar trafficking pathway involves three trafficking routes in Arabidopsis *(i) depending on the sequential action of RAB5 and RAB7, (ii) AP-3-dependent but RAB5- and RAB7- independent pathway, and (iii) RAB5-dependent and AP-3-independent route (modified from Ebine et al. [46]).*

ARF1 manages ER-to-Golgi transport and Golgi-derived transport to the plasma membrane, depending on the COPI vesicle coat protein components [53, 54]. A large number of ARF in plants suggest the possibility for highly regulated vesicle trafficking [53]. Similar to other small GTPases, ARF GTPases cycle between an active state, when associated with GTP (membrane-bound form), and an inactive state when bound to GDP (predominantly cytosolic form).

In its GDP form, ARF1 is present in the cytosol and is recruited to the surface of Golgi membranes by a GEF. A SEC7-type GEF stimulates the binding of GTP to ARF1. This progression can be inhibited by the fungal metabolite Brefeldin A (BFA) in mammalian cells [55]. In the same system, the GDP-bound form of ARF1 interacts with p24 cytosolic tails [56]. The cytosolic ARF1 activation initiates COPI biogenesis. The GTP-bound form of ARF1 interacts with coatomer, which can also interact directly with the p24 cytosolic tails. In this manner, the p24 cytosolic tail can interact both with ARF1 and coatomer [56]. The conformational change of ARF1 occurs by the GTP/GDP exchange that may cause its dissociation from p24 cytosolic tails [56].

COPI vesicles mediate different transport steps, including ER-to-Golgi intermediate compartment transport, Golgi transport, and/or intra-Golgi transport (anterograde transport and/or retrograde transport) [57, 58]. Two types of COPI-coated vesicles form containing anterograde or retrograde cargo (KDEL receptor), and low amounts of Golgi enzymes have been reported to exist at the Golgi apparatus level [59]. COPI proteins are involved in transport along the endocytic pathway [60]. During the selective transport of vesicles, the coat proteins must distinguish between cargo and resident proteins of the donor organelle. *Arabidopsis* has single genes for γ-COP and δ-COP and multiple genes for the other COPI subunits [61]. COPI coatomer forms a coat around vesicles budding from the Golgi. Two different sizes of COPI (COPIa and COPIb) vesicles have been identified by multiparameter electron tomography analysis in *Arabidopsis* [62]. COPIa coats are retrograde transport vesicles, and COPIb vesicles are restricted to medial- and trans-cisternae and are involved for retrograde transport within the Golgi stack. The multiple copies of COPI in plants suggest the presence of different classes of COPI vesicles. The protein complex COPI coatomer is composed of seven subunits (α, β, β', γ, δ, ε, and ζ-COP). COPI represents approximately 0.2% of soluble cytosolic protein indicating their roles as unassembled precursors of COPI vesicles [63].

In intracellular transport, cargo transmembrane protein sorting at each step depends on the specific interaction of certain signals in their cytoplasmic tails with the correct coat proteins [64]. In yeast and mammalian cells, the cytosolic dilysine motif is essential for the ER localization of type I membrane proteins [65]. The two lysine residues must be in the −3, −4 (KKXX) or − 3, −5 (KXKXX) positions relative to the carboxy (C) terminus [65]. For ER localization, the lysine residue at the −3 position is the most critical residue [66]. In mammals, lysine residue mutations within the KKXX motif lead to the expression of reporter proteins at the cell surface [65]. In contrast, the same mutation leads to vacuolar transfer in yeast [67]. The p24 proteins have been suggested to function in Golgi-to-ER retrograde transport, as they contain cargo receptors on their luminal side and coatomer and/or ARF1 receptors on their cytoplasmic side in mammalian cells [68] COPI is necessary for recycling p24 proteins to the ER from the Golgi apparatus [69].

In general p24 proteins are only found in the ER. The binding of p24 proteins to COPI is mediated by dilysine motifs at the −3 and − 4 positions of p24 [69]. Up to 11 different p24 family members proteins have been identified in *Arabidopsis*. The p24 proteins appear to bind COPI with higher affinity than COPII. In the cytosolic tail of the *Arabidopsis* p24 (Atp24), the dilysine motif is important both for binding of coatomer subunits and ARF1 [70].

ARF1 has been shown to localize to Golgi and endosomes and regulate cell proliferation, cell elongation, and fertility, whereas ARF6 is associated with plasma membrane and important for actin remodeling and receptor endocytosis in plant cells [54]. The ARF6 overexpression was shown in breast cancer cells and also involved ERK signaling during invasion. On the other hand, the use of ARF1 protein as a prognostic marker for gastric cancer has been reported [71].

Low expression level of ARF1 (Q71L) mutant in tobacco mesophyll protoplasts has been reported to lead to wtp24 accumulation in the Golgi apparatus [69]. These studies verify the COPI-recycling mechanism can efficiently function in plants. COPI-binding dilysine motif-deficient p24 mutants are transported to the PVC and vacuole [69]. It was observed that p25 may function as an anchor for the p24 proteins in retrograde transport [69].

2.1.2.2 Intra-Golgi transport

Two different models for intra-Golgi transport were suggested: (i) vesicular transport and (ii) cisternal progression/maturation.

Between two different models, the direction of COPI vesicles is a critical distinguishing factor. (i) The *vesicular transport model* assumes that anterograde cargo is transported between static cisternae by coordinated budding and fusion reactions of anterograde-directed COPI vesicles [72]. Retrograde-directed COPI vesicles antagonize the continuous loss of material at the trans-Golgi. Therefore, two different COPI vesicles are involved for this model, one mediating anterograde transport and the other mediating retrograde transport. (ii) In the *cisternal progression/maturation model*, Golgi cisternae are stable compartments. In anterograde COPII vesicles, secretory cargoes are transported from one cisterna to the next, which finally disassemble at the trans-Golgi. Anterograde cargo would not leave the lumen, and resident Golgi proteins are maintained in the cisternae [72].

The COPI vesicles contain Golgi enzymes at a concentration that is up to 10 times higher than that found in the cisternae in animal cells [73]. The cisternal progression/maturation model does not clarify the presence of anterograde cargo within COPI vesicles or different anterograde cargo transportation rates in animal cells [74].

The "cisternal progression/maturation" model is the most widely accepted model for distinct and essential trafficking tasks in the Golgi. The stack of Golgi cisternae involves the historical record of progression from entry at the cis-face to exit at the trans-face [75]. The cargo molecules stay within a given cisternae as it passes, across a regular of seven locations within the Golgi stack on its way to the *trans*-face, and exit from the Golgi by transport carriers. In the cisternal progression, the newly arrived cargo in the Golgi exited with exponential kinetics rather than exhibiting a discrete lag or transit time [76]. Conserved oligomeric Golgi (COG) complex proteins accelerate the tethering of the vesicles to the target cisternae [77]. Resident Golgi proteins are assumed to recycle from older to younger cisternae. In retrograde COPI vesicles, transmembrane Golgi proteins may recycle. Peripheral Golgi proteins may recycle by dissociating from a given cisternae and then bind and combine to a younger cisternae.

2.2 Vacuolar trafficking

Plants have a complex vacuolar transport system different from that of mammalian systems by assigning evolutionarily conserved machinery to unique trafficking pathways. These pathways provide a fundamental basis for plant development at the cellular and higher-ordered levels [78]. Plants have evolved unique and complex vacuolar trafficking pathways compared with non-plant systems (**Figure 3**) [46].

The diverse functions of plant vacuoles are fulfilled through tight regulation of trafficking to and from the vacuoles, which involves evolutionarily conserved machinery components including Rab GTPases [79]. However, the basic framework of the Rab GTPase action is well conserved in eukaryotic cells [80]. Recent comparative genomic studies suggest that each eukaryotic lineage has acquired a unique repertoire of Rab GTPases during evolution [81] Autophagy-related and Golgi-independent transport from the ER to the vacuole is another example of such trafficking pathway [82]. This pathway also involves an exocyst subcomplex, although Rab and SNARE molecules associated with this pathway have not been identified thus far.

2.2.1 RAB GTPases

RAB GTPases constitute the largest family of small GTPases; 57 members are encoded in the *Arabidopsis* genome [83]. Based on their similarity to animal RAB GTPases, RAB GTPases are grouped into eight clades, i.e., RAB1/RABD, RAB2/RABB, RAB5/RABF, RAB6/RABH, RAB7/ RABG, RAB8/RABE, RAB11/RABA, and RAB18/RABC in angiosperms [84]. Plants also harbor a unique set of Rab GTPases partly characterized by diversification of the RAB5 group acting in endosomal/vacuolar trafficking pathways [85]. RAB7 is also proposed to regulate vacuolar traffic in plants [86]. In animal cells, a sequential action of RAB5 and RAB7 mediated by the effector complex HOPS [18] and a guanine nucleotide exchange factor for RAB7 consisting of SAND1/Mon1 and CCZ1 is responsible for the maturation from early to late endosomal compartments (**Figure 4**) [86].

Future studies on the molecular mechanisms of these plant-specific vacuolar trafficking pathways will reveal how plants have used unique vacuolar trafficking routes and how plants have developed their unique vacuolar trafficking pathways during evolution. The tethering of transport vesicles to the target membranes is mediated by the interaction between RAB GTPases and specific sets of tethering factors, many of which have been shown to be RAB effectors, which bind to specific RABs at the GTP-bound active state in yeast and animal systems [80]. After the tethering of transport vesicles by the tethering factors, soluble N-ethylmaleimide-sensitive-factor attachment protein receptors lead to the membrane fusion [80]. Tethering factors comprise long coiled-coil proteins and protein complexes called

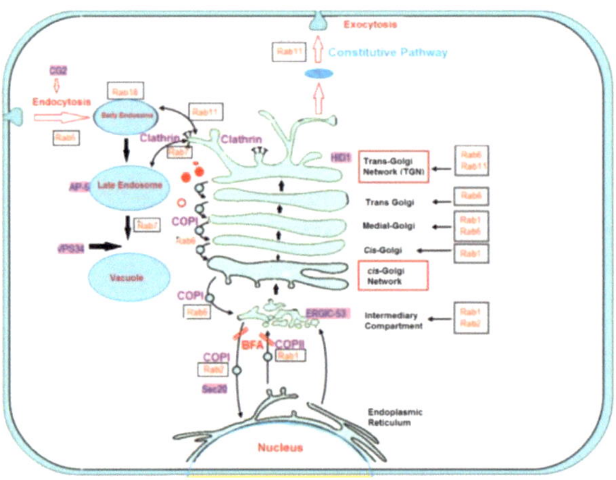

Figure 4.
RAB and the other proteins in intracellular trafficking (modified from Malaria Parasite Metabolic Pathways [47]).

tethering complexes. Each tethering step is mediated by a specific tethering complexes such as COG (functioning in retrograde trafficking within the Golgi), HOPS-CORVET (tethering with the lysosome/vacuole), and exocyst (functioning in the last step of the secretory pathway). In plants, homologous genes encoding these tethering complex proteins are also found, whereas some fibrous coiled-coil proteins are not conserved [84].

2.3 The endocytic pathway

Endocytosis in plant cells has an essential role for basic cellular functions and communication with the environment. Through the formation of closed membrane vesicles (60–120 nm), the uptake of extracellular molecules or the internalization of plasma membrane lipids and proteins is achieved [37]. During the life cycle of the plant, endosomes have vital importance for different processes including lateral organ differentiation, hormone signal transduction, root hair formation, embryo patterning, and plant immunity [36, 87–92]. Transportation of various cargo molecules involved in a broad range of physiological processes from the plasma membrane into the cytoplasm is achieved by this pathway. **Figure 5** shows the general organization of the endosomal trafficking system in plants [36].

As in animal cells, endocytosis in plant cells is mediated by (i) *clathrin-mediated* (CME) and (ii) *clathrin-independent pathways* (CIE) [93].

2.3.1 Clathrin-mediated pathway

In plants, the major endocytic mechanism depends on the coat protein clathrin. This pathway starts at the plasma membrane by clathrin-coated vesicle formation. CME is important for different physiological processes involving cell signaling, cell adhesion, nutrient uptake, developmental regulation, etc. The pathway starts by clathrin-coated vesicle formation at the plasma membrane; in the cytosolic parts of different transmembrane cargo molecules, the clathrin coat binds to specific binding sites. The recruitment of the pioneer proteins to the plasma membrane is ensured by the cargo molecules and enhanced by the initiation of an endocytic pathway. Clathrin involved in a variety of other processes such as the salt stress response, the defense response, cryptogein-induced signaling, and cytokinesis [94–96].

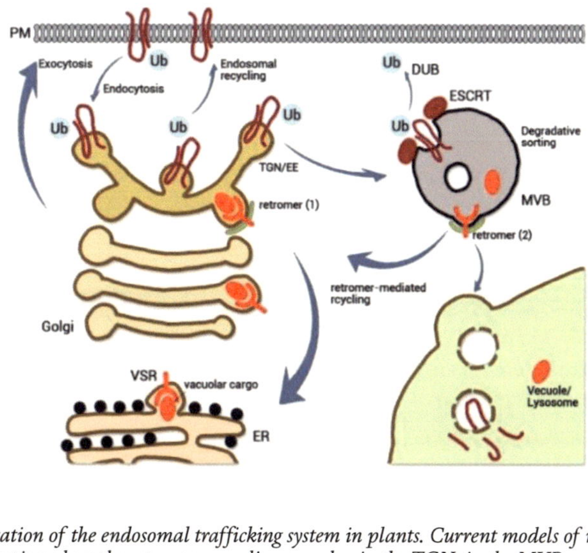

Figure 5.
The general organization of the endosomal trafficking system in plants. Current models of retromer localization and function place the retromer recycling complex in the TGN, in the MVB, or both [36].

Many viruses act as endocytic cargoes for the entry into the cell [97]. A large number of "coat-associated clathrin adaptor proteins" and "scaffold proteins" serve as cargo adaptors by interaction with specific cargoes. Short linear sequence motifs or covalent modifications such as phosphorylation or ubiquitylation in cargo proteins are important for cargo adaptor interactions [98]. Most cargo adaptors also interact directly with lipids and with other coat proteins. These complex interactions cause the initiation of clathrin coat assembly and its further expansion [99]. Binding of the clathrin coat proteins to the cytosolic sites of different transmembrane cargo molecules is essential for the cargo recruitment to the region of the plasma membrane that will form the vesicle.

Because of the difficulty of visualizing and manipulating, studies on cargo and lipids are very difficult [100]. The most widely studied one is the maintenance of polar localization of auxin transporters, namely, the PIN proteins. Polar localization of PIN proteins involves three steps: (i) nonpolar secretion, (ii) clathrin-dependent endocytosis, and (iii) polar recycling [91]. Dynamic regulation of PIN polarity provided with this mechanism is important in response to environmental and developmental stimuli [101, 102]. Mutations in Rab5 or clathrin cause endocytosis disruption and auxin-related developmental defects [91]. Endocytosis works as a negative feedback of the signaling process. After internalization, flagellin-sensitive 2 (FLS2) is targeted for degradation by ubiquitination which then terminates the signal transduction process [103, 104]. The polarly localized *Arabidopsis* boron transporter 1(BOR1), the tomato ethylene-inducing xylanse receptor (LeEIX2), and the iron transporter 1 (IRT1) are the other plasma membrane cargoes in plant cells [92, 105]. For accurate development and growth regulation, the equilibrium of cargo localization in the endomembrane system and the dynamic trafficking machinery is essential during the plant life cycle.

2.3.2 Clathrin-independent pathway

Several endocytic pathways that do not use clathrin-coated vesicles are involved in a CI pathway that was mediated by caveolae. Some of these pathways are constitutive, whereas others are activated by specific signals or by pathogens [106]. Furthermore, their mechanisms and kinetics of endocytic vesicle formation, associated molecular machinery, and cargo direction are different. Some members of the ARF and Rho subfamilies of small GTPases have been suggested to have key roles in regulating different pathways of CI endocytosis [107]. CI pathways are grouped in terms of those that use a "dynamin-mediated scission mechanism" (dynamin-dependent) and those that require other processes (dynamin-independent). A second characteristic is a contribution of small GTPases in several CI pathways [108].

2.4 The endomembrane system in plant development and plant defense

The dysfunction of the endomembrane system can affect plant development and signal transduction [109, 110]. The interaction between the actin cytoskeleton and the endomembrane system involves various aspects of plant cell function and development [111–113].

The actin cytoskeleton is involved for the dynamic feature of the ER [114]. Microtubules have been reported to also influence the mobility of the ER, but to a lesser degree or at a much slower rate [115]. ARF1 plays an essential role in normal cell growth, plant development, and cell polarity and is ubiquitously expressed in all organs of *Arabidopsis* [116, 117]. In de-etiolated pea shoots, ARF1 was concentrated mainly in the crude Golgi fractions [38]. Antisense RNA studies show that ARF also affects cell expansion and cell size in *Arabidopsis* [118]. BFA-visualized exocytic trafficking defective1 (BEX1) has been reported to require for recycling of PIN transporters and auxin-mediated development in *Arabidopsis* [52]. BEX1 encodes ARF1A1C which localizes to the TGN/EE and Golgi apparatus. For normal venation patterning, polar auxin transport by PIN1 is required [118, 119]. Vascular network defective 4 (VAN4) is required for cellular growth and venation development [120]. VAN4 encodes a putative TRS120 subunit of the TRAPPII complex protein that functions as a Rab-GEF and/or tethering factor [121]. VAN4 is involved in polar localization and the recycling of PIN proteins. VAN3/SFC, ARF-GAP, and VAN7/GNOM ARF-GEF have been reported to regulate venation pattern by regulating the activity of the ARF GTPase [122].

Another relation with endomembrane trafficking and plant development was revealed by the continuous vascular ring mutants (COV1). Parker et al. have reported that the COV1 mutant is involved in ectopic differentiation of vascular *Arabidopsis* cells [123]. Afterward, COV1 has been reported as a TGN-localized membrane protein that is required for Golgi morphology and vacuolar protein trafficking and for the development of myrosin cells in leaves [124].

The ubiquitin-proteasome system (UPS) is important for the cytosolic and nuclear protein degradation, whereas certain proteins are degraded by autophagic degradation. De-ubiquitylating enzymes (DUBs) are essential for endosomal trafficking by affecting the fate of endocytosed cargo [125]. Endosomal sorting complexes required for transport (ESCRT) components are crucial for plant growth and development. Mutations of ESCRT or ESCRT-associated proteins in plants lead to ubiquitin accumulation, embryonic and seedling lethality, and misregulation of different signaling pathways, which can be associated with endosomal sorting defects in *Arabidopsis* [10, 126]. ESCRT mutations in *Arabidopsis* cause it to die at different developmental stages [127]. Under optimal growth conditions, autophagy seems to be unessential for plant life cycle. But a lack of autophagy can be the reason of the carcinogenesis and neurodegenerative diseases in the mammalian system [128].

Plants protect themselves with the help of small RNA-dependent immune system in response to biotic stress [129]. sRNAs are short regulatory RNAs (20–30 nucleotides) that silence genes with complementary sequences [130]. Against pathogens, several groups of plant sRNAs have important roles in plant defense. Plants send sRNAs in extracellular vesicles (exosomes) to the pathogen to silence virulence genes [130–132]. Host *Arabidopsis* cells have been shown to secrete exosome-like extracellular vesicles to deliver sRNAs into fungal pathogen *Botrytis cinerea*. These sRNA-containing vesicles accumulate at the infection sites of plant and are occupied up by the fungal cells. Transferred host sRNAs cause silencing of virulence-related genes critical for pathogenicity. Plant extracellular vesicles, mainly exosomes, have been reported to play a crucial role in cross-kingdom sRNA trafficking between *Arabidopsis* and the fungal pathogen *B. cinerea* [129].

3. Conclusions

The functional organization of eukaryotic cells requires the exchange of proteins, lipids, and polysaccharides between membrane compartments through transport intermediates. Transport from one compartment of this pathway to another is mediated by vesicular carriers, which are formed by the controlled assembly of coat protein complexes (COPs) on donor organelles. The plant endomembrane system is mostly conserved among eukaryotes but shows complex features. The structural organization of the endomembrane system is important for correct membrane trafficking and plant physiology. The trans-Golgi network (TGN) is a unique subcellular structure, which is a sorting center that integrates upstream cargoes from secretory vesicles, the plasma membrane, and other organelles. The TGN functions as an early endosome compartment, adding to the complexity of sorting mechanisms in plant cells. Protein sorting at the ER-Golgi interface is important for the protein defects. However, the specificity and quantity of cargo sorting control mechanisms between endosome compartments are not completely clarified. More comprehensive studies on endomembrane trafficking will be necessary for the illumination of development, disease responses, hormone signaling (ABA and auxin), and plant immune system via sRNAs in exosomes in plant cells.

Acknowledgements

This work was supported by grant to B. Cevher-Keskin from ICGEB (CRP/TUR09-03) and COST Action (CA16212)-TUBITAK 2170401. I'm grateful to Prof. Dr. Mahmut TOR for valuable suggestions and Faik Keskin for the preparation of the figures.

Conflict of interest

The author declares no conflict of interest.

Author details

Birsen Cevher-Keskin
Plant Molecular Biology and Genetics Laboratory, Genetic Engineering and Biotechnology Institute, Marmara Research Center, The Scientific and Technological Research Council of Turkey (TUBITAK), Gebze, Kocaeli, Turkey

*Address all correspondence to: bcevherkeskin@gmail.com; birsen.keskin@tubitak.gov.tr

References

[1] Morita MT, Shimada T. The plant endomembrane system-a complex network supporting plant development and physiology. Plant and Cell Physiology. 2014;**55**:667-671. DOI: 10.1093/pcp/pcu049

[2] Inada N, Ueda T. Membrane trafficking pathways and their roles in plant–microbe interactions. Plant and Cell Physiology. 2014;**55**(4):672-686. DOI: 10.1093/pcp/pcu046

[3] Ebine K, Inoue T, Ito J, Ito E, Uemura T, Goh T, et al. Plant vacuolar trafficking occurs through distinctly regulated pathways. Current Biology. 2014;**24**(12):1375-1382. DOI: 10.1016/j.cub.2014.05.004

[4] Inada N, Ueda T. Membrane trafficking pathways and their roles in plant–microbe Interactions. Plant and Cell Physiology. 2014;**55**(4):672-686. DOI: 10.1093/pcp/pcu046

[5] Kahn RA, Der CJ, Bokoch GM. The ras superfamily of GTP-binding proteins: Guidelines on nomenclature. The FASEB Journal. 1992;**6**:2512-2513. DOI: 10.1016/0166-6851(96)02579-0

[6] Vernoud V, Horton AC, Yang Z, Nielsen E. Analysis of the small GTPase gene superfamily of Arabidopsis. Plant Physiology. 2003;**131**:1191-1208. DOI: 10.1104/pp.013052

[7] Inoue H, Randazzo PA. Arf GAPs and their interacting proteins. Traffic. 2007;**8**:1465-1475. DOI: 10.1111/j.1600-0854.2007.00624.x

[8] Bourne HR, Sanders DA, McCormick F. The GTPase superfamily: A conserved switch for diverse cell functions. Nature. 1990;**348**:125-132

[9] Jekely G. Small GTPases and the evolution of the eukaryotic cell. BioEssays. 2003;**25**:1129-1138. DOI: 10.1002/bies.10353

[10] Wang X, Chung KP, Lin W, Jiang L. Protein secretion in plants: Conventional and unconventional pathways and new techniques. Journal of Experimental Botany. 2017;**69**:21-38. DOI: 10.1093/jxb/erx435

[11] Meritxell B, Cutrona MB, Beznoussenko GV, Fusella A, Martella O, Moral P, et al. Silencing of mammalian Sar1 isoforms reveals COPII-independent protein sorting and transport. Traffic. 2013;**14**(6):691-708. DOI: 10.1111/tra.12060

[12] Villarejo A, Burén S, Larsson S, Déjardin A, Monné M, Rudhe C, et al. Evidence for a protein transported through the secretory pathway en route to the higher plant chloroplast. Nature Cell Biology. 2005;7:1224-1231. DOI: 10.1038/ncb1330

[13] Cutrona MB, Beznoussenko GV, Fusella A, Martella O, Moral P, Mironov AA. Silencing of mammalian Sar1 isoforms reveals COPII-independent protein sorting and transport. Traffic. 2013;**14**:691-708. DOI: 10.1111/tra.12060

[14] Ward TH, Brandizzi F. Dynamics of proteins in Golgi membranes: Comparisons between mammalian and plant cells highlighted by photobleaching techniques. Cellular and Molecular Life Sciences. 2004;**61**:172-185. DOI: 10.1007/s00018-003-3355-6

[15] Kang BH, Staehelin LA. ER-to-Golgi transport by COPII vesicles in Arabidopsis involves a ribosome-excluding scaffold that is transferred with the vesicles to the Golgi matrix. Protoplasma. 2008;**234**:51-64. DOI: 10.1007/s00709-008-0015-6

[16] Morishige M, Hashimoto S, Ogawa E, Toda Y, Kotani H, Hirose M, et al. GEP100 links epidermal growth factor receptor signalling to Arf6

activation to induce breast cancer invasion. Nature Cell Biology. 2008;**10**:85-92. DOI: 10.1038/ncb1672

[17] Nakano A, Muramatsu M. A novel GTP-binding protein, Sar1p, is involved in transport from the endoplasmic reticulum to the Golgi apparatus. The Journal of Cell Biology. 1989;**109**:2677-2691. DOI: 10.1083/jcb.109.6.2677

[18] Hanton SL, Chatre L, Matheson LA, Rossi M, Held MA, Brandizzi F. Plant Sar1 isoforms with near-identical protein sequences exhibit different localisations and effects on secretion. Plant Molecular Biology. 2008;**67**:283-294. DOI: 10.1007/s11103-008-9317-5

[19] Orci L, Stamnes M, Ravazzola M, Amherdt M, Perrelet A, Söllner TH, et al. Bidirectional transport by distinct populations of COPI-coated vesicles. Cell. 1997;**90**:335-349

[20] Stephens DJ, Pepperkok R. Illuminating the secretory pathway: When do we need vesicles? Journal of Cell Science. 2001;**114**:1053-1059

[21] D'Enfert C, Wuestehube LJ, Lila T, Schekman R. Sec12p-dependent membrane binding of the small GTP-binding protein Sar1p promotes formation of transport vesicles from the ER. The Journal of Cell Biology. 1991;**114**:663-670. DOI: 10.1083/jcb.114.4.663

[22] Miller EA, Barlowe C. Regulation of coat assembly-sorting things out at the ER. Current Opinion in Cell Biology. 2010;**22**:447-453. DOI: 10.1016/j.ceb.2010.04.003

[23] Miller EA, Antonny B, Hamamoto S, Schekman R. Cargo selection into COPII vesicles is driven by the Sec24p subunit. The EMBO Journal. 2002;**21**:6105-6113. DOI: 10.1093/emboj/cdf605

[24] Miller EA, Beilharz TH, Malkus PN, Lee MCS, Hamamoto S, Orci L, et al. Multiple cargo binding sites on the COPII subunit Sec24p ensure capture of diverse membrane proteins into transport vesicles. Cell. 2003;**114**:497-509. DOI: 10.1016/S0092-8674(03)00609-3

[25] Hughes H, Budnik A, Schmidt K, Palmer KJ, Mantell J, Noakes C, et al. Organisation of human ER-exit sites: Requirements for the localisation of Sec16 to transitional ER. Journal of Cell Science. 2009;**122**:2924-2934. DOI: 10.1242/jcs.044032

[26] Supek F, Madden DT, Hamamoto S, Orci L, Schekman R. Sec16p potentiates the action of COPII proteins to bud transport vesicles. The Journal of Cell Biology. 2002;**158**:1029-1038. DOI: 10.1083/jcb.200207053

[27] Gimeno RE, Espenshade P, Kaiser CA. SED4 encodes a yeast endoplasmic reticulum protein that binds Sec16p and participates in vesicle formation. Journal of Cell Biology. 1995;**131**:325-338. DOI: 10.1083/jcb.131.2.325

[28] Watson P, Townley AK, Koka P, Palmer KJ, Stephens DJ. Sec16 defines endoplasmic reticulum exit sites and is required for secretory cargo export in mammalian cells. Traffic. 2006;**7**:1678-1687. DOI: 10.1111/j.1600-0854.2006.00493.x

[29] De Matteis MA, Luini A. Exiting the Golgi complex. Nature Reviews. Molecular Cell Biology. 2008;**9**:273-284. DOI: 10.1038/nrm2378

[30] Lee MCS, Orci L, Hamamoto S, Futa E, Ravazzola M, Schekman R. Sar1p N-terminal helix initiates membrane curvature and completes the fission of a COPII vesicle. Cell. 2005;**122**:605-617. DOI: 10.1016/j.cell.2005.07.025

[31] Settles EI, Loftus AF, McKeown AN, Parthasarathy R. The vesicle trafficking

protein Sar1 lowers lipid membrane rigidity. Biophysical Journal. 2010;**99**:1539-1545. DOI: 10.1016/j.bpj.2010.06.059

[32] Pagant S, Wu A, Edwards S, Diehl F, Miller EA. Sec24 is a coincidence detector that simultaneously binds two signals to drive ER export. Curr. O Biologico. 2015;**25**:403-412. DOI: 10.1016/j.cub.2014.11.070

[33] Saito-Nakano Y, Nakano A. Sed4p functions as a positive regulator of Sar1p probably through inhibition of the GTPase activation by Sec23p. Genes to Cells. 2000;**5**:1039-1048

[34] Connerly PL, Esaki M, Montegna EA, Strongin DE, Levi S, Soderholm J, et al. Sec16 is a determinant of transitional ER organization. Current Biology. 2005;**15**:1439-1447. DOI: 10.1016/j.cub.2005.06.065

[35] Loftus AF, Hsieh VL, Parthasarathy R. Modulation of membrane rigidity by the human vesicle trafficking proteins Sar1A and Sar1B. Biochemical and Biophysical Research Communications. 2012;**426**:585-589. DOI: 10.1016/j.bbrc.2012.08.131

[36] Reyes FC, Rafael B, Otegui MS. Plant endosomal trafficking pathways. Current Opinion in Plant Biology. 2011;**14**(6):666-673. DOI: 10.1016/j.pbi.2011.07.009

[37] Robinson DG, Herranz MC, Bubeck J, Pepperkok R, Ritzenthaler C. Membrane dynamics in the early secretory pathway. Critical Reviews in Plant Sciences. 2007;**26**:199-225. DOI: 10.1080/07352680701495820

[38] Keskin BC, Yuca E, Ertekin O, Yüksel B, Memon AR. Expression characteristics of ARF1 and SAR1 during development and the de-etiolation process. Plant Biology. 2012;**14**:24-32. DOI: 10.1111/j.1438-8677.2011.00482.x

[39] Fromme JC, Ravazzola M, Hamamoto S, Al-Balwi M, Eyaid W, Boyadjiev SA, et al. The genetic basis of a craniofacial disease provides insight into COPII coat assembly. Developmental Cell. 2007;**13**:623-634. DOI: 10.1016/j.devcel.2007.10.005

[40] Sarmah S, Barallo-Gimeno A, Melville DB, Topczewski J, Solnica-Krezel L, Knapik EW. Sec24D-dependent transport of extracellular matrix proteins is required for zebrafish skeletal morphogenesis. PLoS One. 2010;**5**:e10367. DOI: 10.1371/journal.pone.0010367

[41] Barlowe C. Signals for COPII-dependent export from the ER: What's the ticket out? Trends in Cell Biology. 2003;**13**:295-300. DOI: 10.1016/S0962-8924(03)00082-5

[42] Sieben C, Mikosch M, Brandizzi F, Homann U. Interaction of the K(+)-channel KAT1 with the coat protein complex II coat component Sec24 depends on a di-acidic endoplasmic reticulum export motif. The Plant Journal. 2008;**56**:997-1006. DOI: 10.1111/j.1365-313X.2008.03658.x

[43] Bar-Peled M, Conceicao A, Frigerio L, Raikhel NV. Expression and regulation of aERD2, a gene encoding the KDEL receptor homolog in plants, and other genes encoding proteins involved in ER-Golgi vesicular trafficking. The Plant Cell. 1995;**7**:667-676. DOI: 10.1105/tpc.7.6.667

[44] Marti L, Fornaciari S, Renna L, Stefano G, Brandizzi F. COPII-mediated traffic in plants. Trends in Plant Science. 2010;**15**:522-528. DOI: 10.1016/j.tplants.2010.05.010

[45] Jürgens G. Membrane trafficking in plants. Annual Review of Cell and Developmental Biology. 2004;**20**:481-504. DOI: 10.1146/annurev.cellbio.20.082503.103057

[46] Ebine K, Inoue T, Ito J, Ito E, Uemura T, Goh T, et al. Plant vacuolar trafficking occurs through distinctly regulated pathways. Current Biology. 2014;**24**:1375-1382. DOI: 10.1016/j.cub.2014.05.004

[47] Malaria Parasite Metabolic Pathways. Available from: http://mpmp.huji.ac.il/maps/rabIntracellular.html [Accessed: 01 March 2019]

[48] DaSilva LL, Snapp EL, Denecke J, Lippincott-Schwartz J, Hawes C, Brandizzi F. Endoplasmic reticulum export sites and Golgi bodies behave as single mobile secretory units in plant cells. The Plant Cell. 2004;**16**:1753-1771. DOI: 10.1105/tpc.022673

[49] Jiang L, Patarroyo D, Cabanillas G, Zheng H, Laliberte JF. The vesicle-forming 6K2 protein of turnip mosaic virus interacts with the COPII Coatomer Sec24a for viral systemic infection. Journal of Virology. 2015;**89**(13):6695-6710. DOI: 10.1128/JVI.00503

[50] Deng Y, Srivastava R, Quilichini TD, Dong H, Bao Y, Horner HT, et al. IRE1, a component of the unfolded protein response signaling pathway, protects pollen development in Arabidopsis from heat stress. Plant Journal. 2016;**88**(2):193-204. DOI: 10.1111/tpj.13239

[51] Pimpl P, Hanton SL, Taylor JP, Pinto-daSilva LL, Denecke J. The GTPase ARF1p controls the sequence-specific vacuolar sorting route to the lytic vacuole. The Plant Cell. 2013;**15**:1242-1256. DOI: 10.1105/tpc.010140

[52] Tanaka H, Nodzyłski T, Kitakura S, Feraru MI, Sasabe M, Ishikawa T, et al. BEX1/ARF1A1C is required for BFA-sensitive recycling of PIN auxin transporters and auxin-mediated development in arabidopsis. Plant and Cell Physiology. 2014;**55**(4):737-749. DOI: 10.1093/pcp/pct196

[53] Vernoud V, Horton AC, Yang Z, Nielsen E. Analysis of the small GTPase gene superfamily of Arabidopsis. Plant Physiology. 2003;**131**:1191-1208. DOI: 10.1104/pp.013052

[54] Goldberg J. Structural and functional analysis of the ARF1-ARFGAP complex reveals a role for coatamer in GTP hydrolysis. Cell. 1999;**96**:893-902. DOI: 10.1016/S0092-8674(00)80598-X

[55] Chardin P, Paris S, Antonny B, Robineau S, Béraud-Dufour S, Jackson CL, et al. A human exchange factor for ARF contains Sec7- and pleckstrin-homology domains. Nature. 1996;**384**:481-484

[56] Zhao L, Helms JB, Brugger B, Harter C, Martoglio B, Graf R, et al. Direct and GTP-dependent interaction of ADP ribosylation factor 1 with coatomer subunit beta. Proceedings of the National Academy of Sciences of the United States of America. 1997;**94**:4418-4423. DOI: 10.1073/pnas.94.9.4418

[57] Ostermann J, Orci L, Tani K, Amherdt M, Ravazzola M, Elazar Z, et al. Stepwise assembly of functionally active transport vesicles. Cell. 1993;**75**:1015-1025. DOI: 10.1016/0092-8674(93)90545-2

[58] Beck R, Prinz S, Diestelkotter-Bachert P, Rohling S, Adolf F, Hoehner K, et al. Coatomer and dimeric ADP ribosylation factor 1 promote distinct steps in membrane scission. The Journal of Cell Biology. 2011;**194**:765-777. DOI: 10.1083/jcb.201102095

[59] Orci L, Amherdt M, Ravazzola M, Perrelet A, Rothman JE. Exclusion of Golgi residents from transport vesicles budding from Golgi cisternae in intact cells. The Journal of Cell Biology. 2000;**150**:1263-1269

[60] Whitney JA, Gomez M, Sheff D, Kreis TE, Mellman I. Cytoplasmic coat proteins involved in endosome function. Cell. 1995;**83**:703-713

[61] Sanderfoot A, Raikhel N. The secretory system of Arabidopsis. In: The Arabidopsis Book. Rockville, MD, USA: American Society of Plant Biologists; 2003

[62] Donohoe BS, Kang BH, Staehelin LA. Identification and characterization of COPIa- and COPIb-type vesicle classes associated with plant and algal Golgi. Proceedings of the National Academy of Sciences of the United States of America. 2007;**104**:163-168. DOI: 10.1073/pnas.0609818104

[63] Donaldson JG, Jackson CL. ARF family G proteins and their regulators: Roles in membrane transport, development and disease. Nature Reviews. Molecular Cell Biology. 2011;**12**:362-375. DOI: 10.1038/nrm3117

[64] Aniento F, Helms B, Memon A. How to make a vesicle: Coat protein-membrane interactions. The Golgi apparatus and the plant secretory pathway. Annual Plant Reviews. 2003;**9**:36-62. DOI: 10.1111/j.1600-0854.2008.00791.x

[65] Jackson MR, Nilsson T, Peterson PA. Identification of a consensus motif for retention of transmembrane proteins in the endoplasmic reticulum. The EMBO Journal. 1990;**9**:3153-3162

[66] Hardt B, Bause E. Lysine can be replaced by histidine but not by arginine as the ER retrieval motif for type I membrane proteins. Biochemical and Biophysical Research Communications. 2002;**291**:751-757. DOI: 10.1006/bbrc.2002.6515

[67] Gaynor EC, Heesen S, Graham TR, Aebi M, Emr SD. Signal-mediated retrieval of a membrane protein from the Golgi to the ER in yeast. Journal of Cell Biology. 1994;**127**:653-665. DOI: 10.1083/jcb.127.3.653

[68] Nickel W, Brugger B, Wieland F. Vesicular transport: The core machinery of COPI recruitment and budding. Journal of Cell Science. 2002;**115**:3235-3240

[69] Langhans M, Marcote MJ, Pimpl P, Virgili-López G, Robinson DG, Aniento F. In vivo trafficking and localization of p24 proteins in plant cells. Traffic. 2008;**9**:770-785. DOI: 10.1111/j.1600-0854.2008.00719.x

[70] Contreras I, Ortiz-Zapater E, Aniento F. Sorting signals in the cytosolic tail of membrane proteins involved in the interaction with plant ARF1 and coatomer. The Plant Journal. 2004;**38**:685-698. DOI: 10.1111/j.1365-313X.2004.02075.x

[71] Tsai MM, Lin PY, Cheng WL, Tsai CY, Chi HC, Chen CY, et al. Overexpression of ADP-ribosylation factor 1 in human gastric carcinoma and its clinicopathological. Cancer Science. 2012;**103**:1136-1144. DOI: 10.1111/j.1349-7006.2012.02243.x

[72] Rothman JE, Wieland FT. Protein sorting by transport vesicles. Science. 1996;**272**:227-234. DOI: 10.1126/science.272.5259.227

[73] Lanoix J, Ouwendijk J, Lin CC, Stark A, Love HD, Ostermann J, et al. GTP hydrolysis by arf-1 mediates sorting & concentration of Golgi resident enzymes into functional COP I vesicles. The EMBO Journal. 1999;**18**:4935-4948. DOI: 10.1093/emboj/18.18.4935

[74] Pepperkok R, Whitney JA, Gomez M, Kreis TE. COPI vesicles accumulating in the presence of a GTP restricted arf1 mutant are

depleted of anterograde and retrograde cargo. Journal of Cell Science. 2000;**113**:135-144

[75] Glick BS, Elston T, Oster GA. Cisternal maturation mechanism can explain the asymmetry of the Golgi stack. FEBS Letters. 1997;**414**:177-181. DOI: 10.1016/S0014-5793(97)00984-8

[76] Patterson GH, Hirschberg K, Polishchuk RS, Gerlich D, Phair RD, Lippincott-Schwartz J. Transport through the Golgi apparatus by rapid partitioning within a two-phase membrane system. Cell. 2008;**133**:1055-1067. DOI: 10.1016/j.cell.2008.04.044

[77] Smith RD, Lupashin VV. Role of the conserved oligomeric Golgi (COG) complex in protein glycosylation. Carbohydrate Research. 2008;**343**:2024-2031. DOI: 10.1016/j.carres.2008.01.034

[78] Brillada C, Rojas-Pierce M. Vacuolar trafficking and biogenesis: A maturation in the field. Current Opinion in Plant Biology. 2017;**40**:77-81. DOI: 10.1016/j.pbi.2017.08.005

[79] Wickner W, Schekman R. Membrane fusion. Nature Structural & Molecular Biology. 2008;**15**:658-664

[80] Saito C, Ueda T. Chapter 4: Functions of RAB and SNARE proteins in plant life. International Review of Cell and Molecular Biology. 2009;**274**:183-233. DOI: 10.1016/S1937-6448(08)02004-2

[81] De Marchis F, Bellucci M, Pompa A. Unconventional pathways of secretory plant proteins from the endoplasmic reticulum to the vacuole bypassing the Golgi complex. Plant Signaling & Behavior. 2013;**8**(8):25129. DOI: 10.4161/psb.25129

[82] Kulich I, Pečenková T, Sekereš J, Smetana O, Fendrych M, Foissner I, et al. Arabidopsis exocyst subcomplex containing subunit EXO70B1 is involved in autophagy-related transport to the vacuole. Traffic. 2013;**11**:1155-1165. DOI: 10.1111/tra.12101

[83] Hill D, Sylvester A. Diversification of the *Rab* guanosine triphosphatase family in dicots and monocots. Journal of Integrative Plant Biology. 2007;**49**:1129-1141. DOI: 10.1111/j.1672-9072.2007.00520.x

[84] Fujimoto M, Ueda T. Conserved and plant-unique mechanisms regulating plant post-Golgi traffic. Frontiers in Plant Science. 2012;**3**:197. DOI: 10.3389/fpls.2012.00197

[85] Viotti C, Kruger F, Krebs M, Neubert C, Fink F, Lupanga U, et al. The endoplasmic reticulum is the main membrane source for biogenesis of the lytic vacuole in Arabidopsis. Plant Cell. 2013;**25**:3434-3449. DOI: 10.1105/tpc.113.114827

[86] Bock JB, Matern HT, Peden AA, Scheller RH. A genomic perspective on membrane compartment organization. Nature. 2001;**409**:839-841. DOI: 10.1038/35057024

[87] Carter CJ, Bednarek SY, Raikhel NV. Membrane trafficking in plants: New discoveries and approaches. Current Opinion in Plant Biology. 2004;**7**:701-707. DOI: 10.1016/j.pbi.2004.09.016

[88] Voigt B, Timmers AC, Šamaj J, Hlavačka A, Ueda T, Preuss M, et al. Actin-based motility of endosomes is linked to the polar tip growth of root hairs. European Journal of Cell Biology. 2005;**84**:609-621. DOI: 10.1016/j.ejcb.2004.12.029

[89] Robatzek S, Chinchilla D, Boller T. Ligand-induced endocytosis of the pattern recognition receptor FLS2 in Arabidopsis. Genes Development. 2006;**20**:537-542. DOI: 10.1101/gad.366506

[90] Geldner N, Hyman DL, Wang X, Schumacher K. Chory J Endosomal signaling of plant steroid receptor kinase BRI1. Genes Development. 2007;**21**:1598-1602. DOI: 10.1101/gad.1561307

[91] Dhonukshe P, Tanaka H, Goh T, Ebine K, Mahonen AP, Prasad K, et al. Generation of cell polarity in plants links endocytosis, auxin distribution and cell fate decisions. Nature. 2008;**456**:962-966. DOI: 10.1038/nature07409

[92] Sharfman M, Bar M, Ehrlich M, Schuster S, Melech-Bonfil S, Ezer R, et al. Endosomal signaling of the tomato leucine-rich repeat receptor-like protein LeEix2. The Plant Journal for Cell and Molecular Biology. 2011;**68**:413-423. DOI: 10.1111/j.1365-313X.2011.04696.x

[93] Murphy AS, Bandyopadhyay A, Holstein SE, Peer WA. Endocytotic cycling of PM proteins. Annual Review of Plant Biology. 2005;**56**:221-251. DOI: 10.1146/annurev.arplant.56.032604.144150

[94] König S, Ischebeck T, Lerche J, Stenzel I, Heilmann I. Salt-stress induced association of phosphatidylinositol 4,5-bisphosphate with clathrin-coated vesicles in plants. Biochemical Journal. 2008;**415**:387-399. DOI: 10.1042/BJ20081306

[95] Leborgne-Castel N, Lherminier J, Der C, Fromentin J, Houot V, Simon-Plas F. The plant defense elicitor cryptogein stimulates clathrin-mediated endocytosis correlated with reactive oxygen species production in bright yellow-2 tobacco cells. Plant Physiology. 2008;**146**:1255-1266. DOI: 10.1016/j.funbio.2015.09.011

[96] Karahara I, Suda J, Tahara H, Yokota E, Shimmen T, Misaki K, et al. The preprophase band is a localized center of clathrin-mediated endocytosis in late prophase cells of the onion cotyledon epidermis. Plant Journal. 2009;**57**:819-831. DOI: 10.1111/j.1365-313X.2008.03725.x

[97] Sigismund S et al. Endocytosis and signaling: Cell logistics shape the eukaryotic cell plan. Physiological Reviews. 2012;**92**:273-366. DOI: 10.1152/physrev.00005.2011

[98] Liu AP, Aguet F, Danuser G, Schmid SL. Local clustering of transferrin receptors promotes clathrin-coated pit initiation. Journal of Cell Biology. 2010;**191**:1381-1393. DOI: 10.1083/jcb.201008117

[99] Traub LM, Bonifacino JS. Cargo recognition in clathrin-mediated endocytosis. Cold Spring Harbor Perspectives in Biology. 2013;**5**:a016790

[100] Liu AP, Aguet F, Danuser G, Schmid SL. Local clustering of transferrin receptors promotes clathrin-coated pit initiation. Journal of Cell Biology. 2010;**191**:1381-1393. DOI: 10.1083/jcb.201008117

[101] Kleine-Vehn J, Ding Z, Jones A, Tasaka M, Morita M, Friml J. Gravity-induced PIN transcytosis for polarization of auxin fluxes in gravity-sensing root cells. Proceedings of the National Academy of Sciences of the United States of America. 2010;**21**:22344-22349. DOI: 10.1073/pnas.1013145107

[102] Rakusová H, Gallego-Bartolome J, Vanstraelen M, Robert HS, Alabadi D, et al. Polarization of PIN3-dependent auxin transport for hypocotyl gravitropic response in Arabidopsis thaliana. Plant Journal. 2011;**67**:817-826. DOI: 10.1111/j.1365-313X.2011.04636.x

[103] Göhre V, Spallek T, Häweker H, Mersmann S, Mentzel T, Boller T, et al. Plant pattern-recognition receptor FLS2 is directed for degradation by

the bacterial ubiquitin ligase AvrPtoB. Current Biology. 2008;**18**:1824-1832. DOI: 10.1016/j.cub.2008.10.063

[104] Wada M, Ludewig U, Schaaf G, von Wiren N, Fujiwara T. The Arabidopsis major intrinsic protein NIP5;1 is essential for efficient boron uptake and plant development under boron limitation. The Plant Cell. 2006;**18**:1498-1509. DOI: 10.1105/tpc.106.041640

[105] Barberon M, Dubeaux G, Kolb C, Isono E, Zelazny E, Vert G. Polarization of iron-regulated transporter 1 (IRT1) to the plant-soil interface plays crucial role in metal homeostasis. PNAS. 2014;**111**(22):8293-8298. DOI: 10.1073/pnas.1402262111

[106] Bitsikas V, Correa IR Jr, Nichols BJ. Clathrin-independent pathways do not contribute significantly to endocytic flux. eLife. 2014;**3**:e03970. DOI: 10.7554/eLife.03970

[107] Mayor S, Parton RG, Donaldson JG. Clathrin-independent pathways of endocytosis. Cold Spring Harbor Perspectives in Biology. 2014;**6**(6):a016758. DOI: 10.1101/cshperspect.a016758

[108] Mayor S, Pagano RE. Pathways of clathrin-independent endocytosis. Nature Reviews. Molecular Cell Biology. 2007;**8**(8):603-612. DOI: 10.1038/nrm2216

[109] Ding Y, Robinson DG, Jiang L. Unconventional protein secretion (UPS) pathways in plants. Current Opinion in Cell Biology. 2014;**29**:107-115. DOI: 10.1016/j.ceb.2014.05.008

[110] Boutté Y, Moreau P. Modulation of endomembranes morphodynamics in the secretory/retrograde pathways depends on lipid diversity. Current Opinion in Plant Biology. 2014;**22**:22-29. DOI: 10.1016/j.pbi.2014.08.004

[111] Sampathkumar A, Gutierrez R, Mcfarlane HE, Bringmann M, Lindeboom J, Emons AM, et al. Patterning and lifetime of plasma membrane-localized cellulose synthase is dependent on actin organization in Arabidopsis interphase cells. Plant Physiology. 2013;**162**:675-688. DOI: 10.1104/pp.113.215277

[112] Wang P, Hussey PJ. Interactions between plant endomembrane systems and the actin cytoskeleton. Frontiers in Plant Science. 2015;**6**:422. DOI: 10.3389/fpls.2015.00422

[113] Cevher-Keskin B. ARF1 and SAR1 GTPases in endomembrane trafficking in plants. International Journal of Molecular Sciences. 2013;**14**:18181-18199. DOI: 10.3390/ijms140918181

[114] Sparkes I, Hawes C, Frigerio L. FrontiERs: Movers and shapers of the higher plant cortical endoplasmic reticulum. Current Opinion in Plant Biology. 2011;**14**:658-665. DOI: 10.1016/j.pbi.2011.07.006

[115] Hamada T, Ueda H, Kawase T, Hara-Nishimura I. Microtubules contribute to tubule elongation and anchoring of endoplasmic reticulum, resulting in high network complexity in Arabidopsis. Plant Physiology. 2014;**166**:1869-1876. DOI: 10.1104/pp.114.252320

[116] Kahn RA, Gilman AG. The protein cofactor necessary for ADP-ribosylation of Gs by cholera toxin is itself a GTP binding protein. The Journal of Biological Chemistry. 1986;**261**:7906-7911

[117] Stearns T, Willingham MC, Botstein D, Kahn RA. ADP-ribosylation factor is functionally and physically associated with the Golgi complex. PNAS. 1990;**87**(3):1238-1242. DOI: 10.1073/pnas.87.3.1238

[118] Gebbie LK, Burn JE, Hocart CH, Williamson RE. Genes encoding ADP-ribosylation factors in *Arabidopsis thaliana* L. Heyn.; genome analysis and antisense suppression. Journal of Experimental Botany. 2005;**56**:1079-1091. DOI: 10.1093/jxb/eri099

[119] Zhang J, Nodzyński T, Pěnčík A, Rolčík J, Friml J. PIN phosphorylation is sufficient to mediate PIN polarity and direct auxin transport. PNAS. 2010;**107**(2):918-922. DOI: 10.1073/pnas.0909460107

[120] Naramoto S, Otegui MS, Kutsuna N, de Rycke R, Dainobu T, Karampelias M, et al. Insights into the localization and function of the membrane trafficking regulator GNOM ARF-GEF at the Golgi apparatus in Arabidopsis. The Plant Cell. 2014;**26**:3062-3076. DOI: 10.1105/tpc.114.125880

[121] Naramoto S, Nodzyński T, Dainobu T, Takatsuka H, Okada T, Friml J, et al. VAN4 encodes a putative TRS120 that is required for normal cell growth and vein development in Arabidopsis. Plant and Cell Physiology. 2014;**55**(4):750-763. DOI: 10.1093/pcp/pcu012

[122] Naramoto S, Sawa S, Koizumi K, Uemura T, Ueda T, Friml J, et al. Phosphoinositide dependent regulation of VAN3 ARF-GAP localization and activity essential for vascular tissue continuity in plants. Development. 2009;**136**:1529-1538. DOI: 10.1242/dev.030098

[123] Parker G, Schofield R, Sundberg B, Turner S. Isolation of COV1, a gene involved in the regulation of vascular patterning in the stem of Arabidopsis. Development. 2003;**130**:2139-2148. DOI: 10.1242/dev.00441

[124] Shirakawa M, Ueda H, Koumoto Y, Fuji K, Nishiyama C, Kohchi T, et al. Continuous vascular ring (COV1) is a trans-Golgi network-localized membrane protein required for Golgi morphology and vacuolar protein sorting. Plant and Cell Physiology. 2014;**55**(4):764-772. DOI: 10.1093/pcp/pct195

[125] Kalinowska K, Isono E. All roads lead to the vacuole-autophagic transport as part of the endomembrane trafficking network in plants. Journal of Experimental Botany. 2018;**69**:1313-1324. DOI: 10.1093/jxb/erx395

[126] Nagel M-K, Kalinowska K, Vogel K, Reynolds GD, Wu Z, Anzenberger F, et al. Arabidopsis SH3P2 is an ubiquitin-binding protein that functions together with ESCRT-I and the deubiquitylating enzyme AMSH3. PNAS. 2017;**114**(34):7197-7204. DOI: 10.1073/pnas.1710866114

[127] Kolb C, Nagel M-K, Kalinowska K, Hagmann J, Ichikawa M, Anzenberger F, et al. FYVE1 is essential for vacuole biogenesis and intracellular trafficking in *Arabidopsis thaliana*. Plant Physiology 2015;**67**:1361-1373. DOI: 10.1104/pp.114.253377

[128] Rusten TE, Simonsen A. ESCRT functions in autophagy and associated disease. Cell Cycle. 2008;**7**(9):1166-1172. DOI: 10.4161/cc.7.9.5784

[129] Zhang C, Wu Z, Li Y, Wu J. Biogenesis, function, and applications of virus-derived small RNAs in plants. Frontiers in Microbiology. 2015;**6**:1237

[130] Baulcombe D. RNA silencing in plants. Nature. 2004;**431**:356-363. DOI: 10.1038/nature02874

[131] Cai Q, Qiao L, Wang M, He B, Lin F-M, Palmquist J, et al. Plants send small RNAs in extracellular vesicles to fungal pathogen to silence virulence genes. Science. 2018;**360**:1126-1129. DOI: 10.1126/science.aar4142

[132] Rovenich H, Boshoven JC, PHJ B, Thomma BP. Filamentous pathogen effector functions: Of pathogens, hosts and microbiomes and microbiomes. Current Opinion in Plant Biology. 2014;**20**:96-103. DOI: 10.1016/j.pbi.2014.05.001

Chapter 3

Electrolytic Cell Applied for the Breakdown of Endocrine Disrupting Drugs in Aqueous Tributaries

Jorge Alberto Mendoza Pérez,
Abril Gardenia Martínez Castillo,
Jorge Octaviano Gomez Castrejon
and Juan Carlos Gómez Buendía

Abstract

In this chapter, we report previous results about advances of an electrolysis process developed for breakdown of endocrine disrupting drugs in aqueous media. The objective is to achieve the breakdown of two drugs: trimethoprim and a mixture of clavulanic acid-amoxicillin (1:7) with an electrolytic cell by means of oxidization-reduction reactions. The evaluation of the process was carried out using spectrometry techniques UV-Vis, thin layer chromatography (TLC), chemical oxygen demand (COD), and total organic carbon (TOC). Handcrafted mineral carbon electrodes doped with titanium dioxide were designed, platinum and copper wires were placed, and a potassium hydroxide solution was used as electrolyte. The electrolyte, being an alkaline salt, allows the transport of charges from one side to the other, and electrode doped with titanium dioxide is used in order to help the electronic transfer, and the mineral carbon, having a strong affinity for organic and non-polar compounds, performs an adsorption process. Results from several performed assays showed that after 1 hour of treatment, it can be seen the breakdown of the drugs present in a synthetic wastewater solution.

Keywords: electrolysis, oxidation, reduction, drug, depuration, breakdown, absorbance, concentration, electrode

1. Introduction

Unregulated emerging pollutants enter aquatic systems through wastewater treatment plants after consumption and use by humans and animals [1].

This poses a significant risk to aquatic organisms and to public health. Among the main effects described to date are the appearance of changes in fish reproduction due to the presence of hormones and inhibition of photosynthesis in algae

 IntechOpen

by the presence of these [2]. We know that the presence of antibiotics in the environment can make bacteria in wastewater relatively more resistant to them and resistant microorganisms develop.

To make clear how these molecules act as endocrine disruptors, we must indicate that they interfere with the body's homeostasis, usually by mimicking the natural hormones that lead to the activation or blocking of their receptors [3].

Trimethoprim in Mexico is one of the most used drugs for treating urinary tract infections, and it is commonly used in the foreign tourism for attending the traveler's diarrhea [4]. Trimethoprim is incompletely metabolized by humans during the therapeutic process and approximately 80% is excreted in the pharmacologically active form, which can promote the development of bacterial resistance to this compound's form promoting it as an emerging contaminant [5].

Clavulanic acid-amoxicillin is a mixture of two drugs in typical commercial compositions of 185/125 and 500/125 mg and commercially known in México as Augmentin and Clamoxin or Gimaclav, respectively. It is indicated for the treatment of acute and chronic infections of the upper and lower respiratory tracts, meningitis and genitourinary, skin, soft tissue, gastrointestinal and biliary infections, and in general for the treatment of infections caused by pathogens sensitive to this mixture of drugs [6].

Both drugs have been studied by our research group as emerging pollutants with an effect of endocrine disruptors due to their high presence in wastewater of domestic and hospital effluents in México (work in process of being published) and their consequent impact on all types of aquifers, which could act as final receiving bodies. However, in our research group, different advanced oxidation processes have been used and improved for the removal of these kinds of contaminants, these processes include cavitation, photo-catalytic oxidation, or Fenton chemistry, but they have high costs. New expectations were found when it has been applied an electrochemical process.

Electrochemical oxidation is based on the application of an electric current or a potential difference between two electrodes (anode and cathode), wherewith hydroxyl radicals or other oxidizing species can be generated, depending on the anode material used and the type of electrolyte support used [7].

In this chapter, the use of electrolysis for the removal of pharmaceutical-type pollutants is based on the chemical reactions that are carried out between the electrodes submerged in electrolytic solutions by effect of the passage of the electric current, being the function of the electrolyte to serve as a means of transporting electrical loads and provoke the reactions of oxidation-reduction for the degradation of compounds in order to transform them into less hazardous compounds for the environment. The electrolyte being of the family of salts allows the anions to carry negative loads toward the anode and the cations transport the positive loads to the cathode [8]; the effectiveness of several visible light-activated TiO_2 photocatalysts has been proven for the treatment of emerging contaminants. Doping or co-doping of titanium dioxide using nitrogen, nitrogen-silver, sulfur, carbon, and copper and incorporating graphene nano-leaves increases its effectiveness. The use of titanium dioxide is to improve photocatalytic activity [9]. Considering all the above backgrounds, the experiments reported hereby were performed in order to establish if the proposed system could breakdown or transform the pollutants (drugs) into simpler molecules.

Based on different references, it is known that the trimethoprim can be determined with an absorbance at 237.6 nm [10] or that it can be displayed at a wavelength of 283–350 nm [11], and it is mentioned that in visible light, it is still observed at a wavelength between 400 and 600 nm [12]. In the case of amoxicillin-clavulanic acid mixture, the spectra recorded are in a wavelength range of 200–380 nm [13]. Therefore, the results on the UV-Vis spectrum were analyzed within these intervals.

2. Methodology and materials

2.1 Preparing the model pollutant

For the preparation of synthetic wastewater (SWW) contaminated by trimethoprim and clavulanic acid-amoxicillin, three samples with 500 mL of clean water (0.1 mS/cm) were added with different drug concentrations (trimethoprim 0.1%, trimethoprim 0.2%, and clavulanic acid-amoxicillin 0.2%). Type curves for each drug were prepared with 2 g/L (0.2%) as the highest concentration and 0.1 g/L (0.01%) as the lowest concentration.

The drugs supplied by Merck | Sigma-Aldrich were white powders, and the solutions prepared were kept in refrigeration (6°C) until needed.

2.2 Electrode construction

For the construction of the electrodes, two glass tubes were used, each one of 6.1 cm length and 0.7 cm width, containing the following elements: an electrode connected to the negative pole (cathode) was handcrafted with mineral carbon introducing a copper wire (0.3 mm Ø) and the electrode connected to the positive pole (anode) was also handcrafted with mineral carbon doped with TiO_2 and it was used a platinum wire (0.1 mm Ø). In the cathode, the reduction processes take place and oxidation occurs in the anode, where both are connected to a power supply with a 5 V electrical potential (**Figure 1**).

Platinum is an inert metal that does not participate in the redox reaction but only exchanges electrons [14] since it has a low reactivity with oxygen and with water [15]. Copper is a metal with great ductility and high electrical conductivity [16], which leads to the conclusion that both metal wires are good conductors and will show little corrosion. Furthermore, platinum has catalytic activity and low reactivity. Due to these characteristics, the platinum and copper wires were selected for the electrodes. Doping carbon with TiO_2 aids a better oxidation process at the anode, in addition to having photocatalytic properties that can improve the dissociation reaction in the electrolytic cell.

Figure 1.
TiO_2-doped electrodes and coal.

2.2.1 Obtaining mineral coal

The mineral coal was obtained from zinc-carbon battery carbon cylinders free of cadmium (0% Cd) and (0% Mg) magnesium, confirmed by atomic absorption; so, the mineral coal is free of these elements. Extracted amounts were obtained from different brands of batteries (MARKS rocket, explosive, Panasonic super hyper, King Kong, Sony-new ultra, and Kodak extra heavy-duty, among others). The cylinders were washed by immersion overnight (almost 8 hours) and rinsed with deionized water type I. The zinc and manganese content were also analyzed by atomic absorption of all the carbon cylinders obtained; determination of ammonium chloride was also performed but in the remaining rinse water used to clean the cylinders (the chlorides were quantified by the Mohr method and ammonia by the selective electrode method). Curiously after washing and rinsing the carbon cylinders, the concentrations of zinc and ammonium chloride were not detectable and there is only less than 0.1% impregnation of manganese, which was considered negligible for the experiment.

The coal cylinders were divided into two portions, one was ground directly and the other portion was doped with TiO_2 and ground.

2.2.2 TiO_2-doped coal

The mineral carbon was added into a TiO_2 solution at 3.46 g/L (3.46%) and after that it was introduced to a stove at 100°C for approximately 2 hours. Finally, the mineral carbon doped with TiO_2 was crushed, until a homogenous fine powder (0.65–1 μm) was obtained.

2.2.3 Coal electrodes

Having prepared both coals (doped and not doped), the glass cylinders were filled separately with them and the copper and platinum wires were placed along leaving 7.5 cm of free wire.

2.3 Installation of the electrolytic cell

To assemble the electrochemical cell, it was designed with an acrylic cell, a power supply (5 V DC) from STEREN, a porous rectangle glass membrane (length 8.5 cm, height 3.8 cm, and width 7.0 mm), and the handcrafted electrodes. For the proper functioning of the electrolyte cell, moving of charges from the anode to the cathode and to balance the aqueous solution, a KOH electrolyte (0.01 M) from Merck | Sigma-Aldrich was used. The experiments were performed using 50 mL of the mentioned electrolyte solution and about 200–240 mL of SWW, inside the cell. All the experiments were performed in triplicate.

2.4 Experimental sampling

Twenty milliliters of samples of treated synthetic wastewater (TSWW) were taken every half an hour for an experimentation time period between 1.5 and 2 hours, and at time 0, intermediate and final time samples were collected for each experiment. Parameters such as pH, conductivity, and temperature for each sample were measured. A Hach brand Pro HR pocket conductivity tester was used; the temperature is given in °C and the conductivity in mS/cm.

Determination of total organic carbon (TOC) was performed for all the samples using a Thermo Scientific HiperTOC using the US standard US EPA 415.3.

Another analytical method performed also for all the samples was the chemical oxygen demand (COD) under the Mexican Standard NMX-AA-030/2-SCFI-2011. Both methods were carried out by an IPN reference laboratory, which is certified by the Mexican Accreditation Entity (MAE). Also, each sample was taken for analyzing with thin layer chromatography (TLC) and UV-Vis spectrometry analysis at different wavelengths.

2.5 Analytical method of UV-Vis spectrometry

A visible ultraviolet spectrometer BECKMAN brand model DU 7500i was used for the UV-Vis spectrometry technique. First, an absorbance sweep was performed with the SWW samples at different concentrations, in order to find the highest absorbance wavelength to perform their type curves for both drugs. Subsequently, each of the TSWW samples was also measured at a wavelength ranging from 200 to 350 nm (UV absorbance) and 400 to 800 nm (Vis absorbance) in order to observe the degree of degradation or breakage that was achieved.

In order to validate the analytical methodology, the technical guide on traceability and uncertainty is used with those analytical measurements performed with an ultraviolet-visible spectrophotometry technique, which supports the application of the NMX-17025-IMNC-2006 standard, the foregoing established by the National Metrology Center (NMC) and the Mexican Accreditation Entity (MAE) [17].

2.6 TLC analytical method

For thin-layer chromatography, a 2.5 cm by 4 cm thin gel plate was used as a stationary phase; also, both drug standards from Merck | Sigma-Aldrich, two glass chromatographic cameras, and a 254 nm wavelength UV frame were needed. Polar and non-polar (intermediate) solutions were used as mobile phases (eluents). The polar and non-polar (intermediate) eluent solutions were prepared with different concentrations of hexane and ethyl acetate, both from chromatographic grade (Merck | Sigma-Aldrich). The non-polar (intermediate) contained a 4:6 ratio of ethyl acetate and hexane. The polar eluent does a 6:4 ratio of ethyl acetate and hexane. The TSWW samples were mixed with ethyl acetate in a 4:3 ratio inside a test tube and agitated for 30 seconds.

From the test tube, an aliquot of the TSWW samples was taken with a capillary. For each experiment, the TSWW samples were taken at the start, intermediate, and final times, the aliquots were placed subsequently on the thin plate, and the plate was introduced to the chromatographic chambers each with a different eluent (polar and non-polar). After taken out from the camera, plates were revealed under the 254 nm ultraviolet light. The running fronts for standards and sample components were calculated. The intention to use this method is because it is applied in the pharmaceutical industry as a compound purity determination. Therefore, with this technique, it has been able to see qualitatively whether there is a decrease or breaking of the contaminants in the TSWW sample compared against a SWW sample.

3. Results and discussions

The model designed and constructed for the electrolytic cell used in the breakdown of the trimethoprim and acid clavulanic-amoxicillin is described in **Figure 2**.

The type curves for both drugs that were obtained and are showed in **Figures 3** and **4** with their corresponding equations and linear coefficient.

$$Y = 0.1959x + 1.4541 \tag{1}$$

$$R^2 = 0.9782 \tag{2}$$

Eq. (1) shows the linearity, for the trimethoprim.

$$Y = 0.3535x + 0.004 \tag{3}$$

$$R^2 = 0.9612 \tag{4}$$

Equation (3) shows the linearity for the acid clavulanic.

The results obtained from the UV-Vis spectrometry technique at 350 nm is shown in **Figure 5** and results with a 400 nm wavelength is shown in **Figure 6**, where the absorbance units were plotted with respect to the treatment time. When

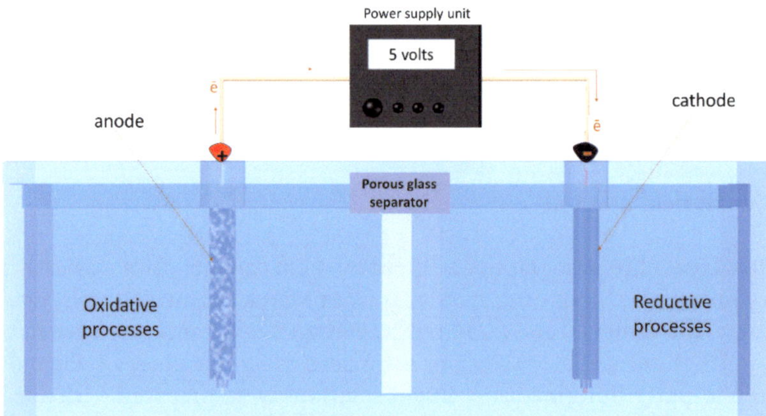

Figure 2.
Main components of an electrolytic cell.

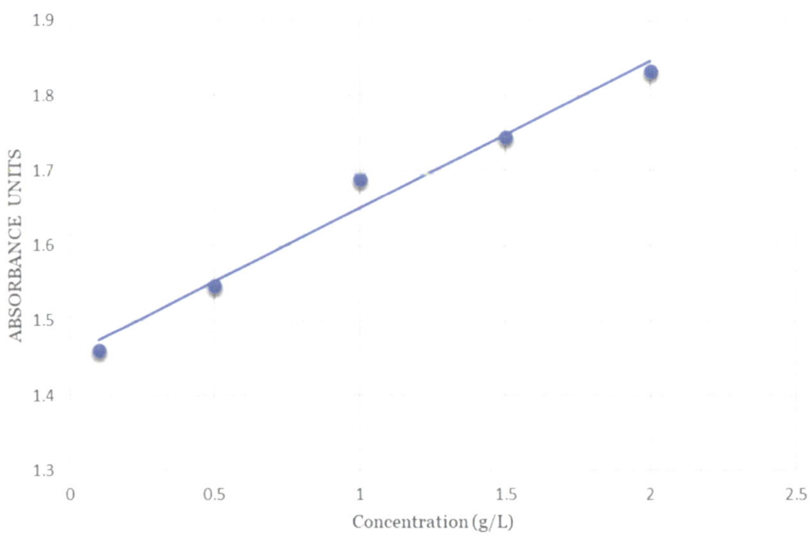

Figure 3.
Type curve of trimethoprim at a wavelength of 300 nm.

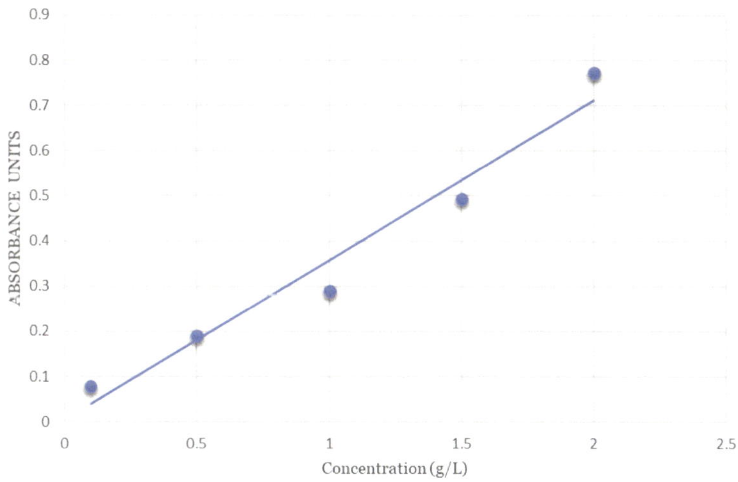

Figure 4.
Clavulanic acid-amoxicillin type curve with a wavelength of 300 nm.

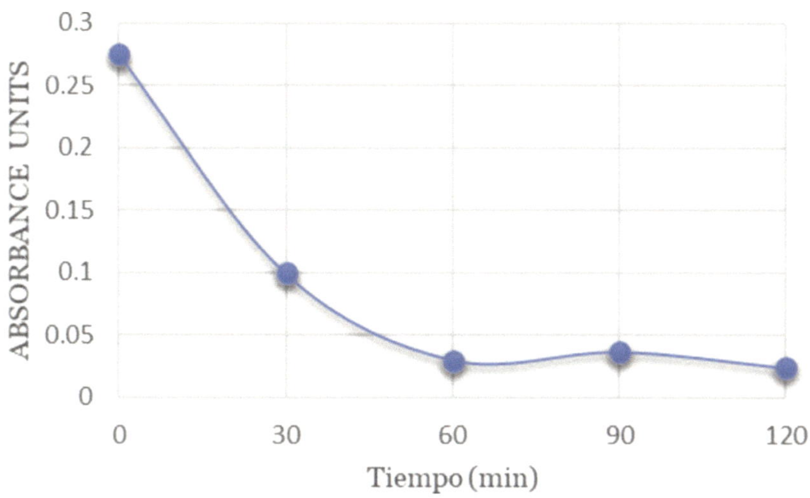

Figure 5.
Mean absorbance decreasing of trimethoprim at different treatment times, at a wavelength of 350 nm.

handling a compound that has rings and double bonds in its molecular structure, it is able to absorb ultraviolet light, and according to the Lambert-Beer law, the absorbance of a substance is proportional to the concentration of the compound.

From the graphs above, we can observe a decay of the absorbance in the trimethoprim at 90 minutes and an increase in the clavulanic acid-amoxicillin mixture experiment after 30 minutes of treatment. These results indicate that there is surely a breakdown in both molecules, but the structure of the molar fragments for trimethoprim differs with respect to the clavulanic acid-amoxicillin mixture. In the first one, the absorbance decreases, but in the second, it increases. It is known that the molecules whose absorbance decays in an electrochemical process present a process of chemical reduction (electron gain), while the compounds whose absorbance increases are understood to undergo chemical oxidation processes (loss of electrons).

For the clavulanic acid-amoxicillin mixture, the increase in absorbance may be due to structural changes in the molecule, so the chemical structure was modified

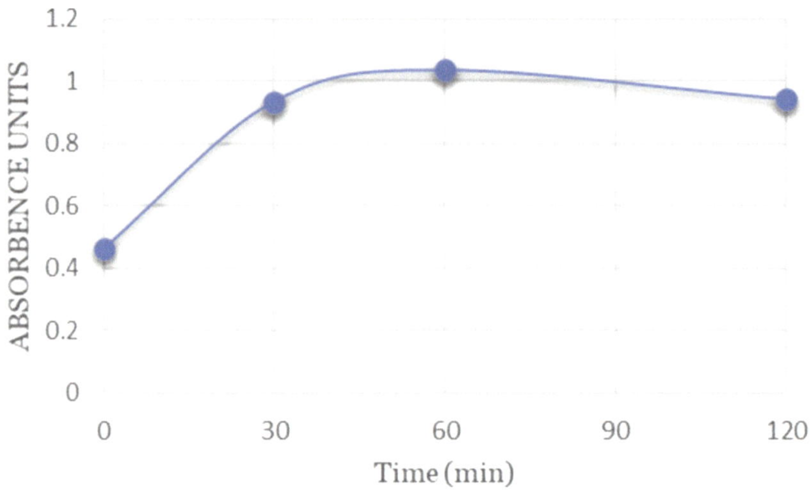

Figure 6.
Absorbance units of the clavulanic acid-amoxicillin mixture at different treatment times, at a wavelength of 400 nm.

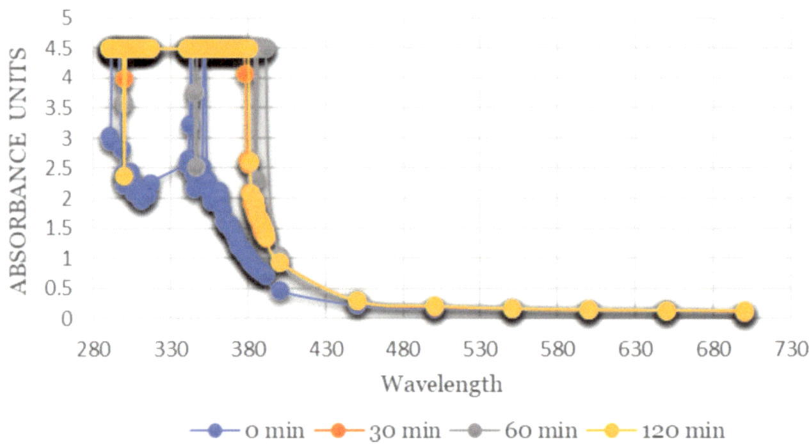

Figure 7.
Full scan of the UV-Vis spectrometer for samples at different treatment times of the clavulanic acid-amoxicillin mixture.

during the reaction, generating resonance systems that will show higher absorbances. Furthermore, this increase in absorbance may be due to the formation of other compounds in the electrolysis reaction, such as hydrogen near the cathode and oxygen near the anode. In the disintegration of water, both ozone and hydrogen peroxide are formed in small quantities [18]; it is known that ozone strongly absorbs radiation in the infrared, visible, and ultraviolet regions [19]. The maximum absorption occurs at a wavelength of 253.7 nm [19], which can be observed in the scanning of the spectrophotometer in **Figure** 7. For these experiments, the wavelength was taken at 400 nm because the sample presented a light yellow hue coloration, which may be due to the fact that the spectral properties of organic molecules depend on the type of valence electrons, on their quantum possibilities of absorption of UV-Vis radiation, and on the presence of chromophoric groups in their structures [20]; therefore, the sigma (σ) electrons make up the single saturated bonds or molecular bonding orbitals of the sigma type, while the pi (π) electrons make up the multiple unsaturated bonds or pi orbitals. These unsaturated groups are called chromophoric groups [21]. The appearance of color in some organic substances is related to the presence of one or more chromophoric groups whose pi electrons are easily excited by the absorption of radiation from the near and visible ultraviolet region (200–800 nm), of corresponding energy (specific length) to the quantum possibilities for the electronic transition [21]. Therefore, it is not certain that in the case of the clavulanic acid-amoxicillin mixture, there is an adequate decomposition, but for the trimethoprim component, this degradation is observed.

The electrical conductivity is in correlation of the amount of ionizable molecules or radicals present in the water as shown in **Figures 8** and **9**; as time passes, conductivity increases, meaning that the drug molecules are being broken and more ionizable fragments are generated.

The COD graph that was obtained from the Trimethoprim TSWW samples showed a tendency to decrease according to the time elapsed. The initial COD at time 0 for the first experiment (trimethoprim 0.2%) had a concentration of

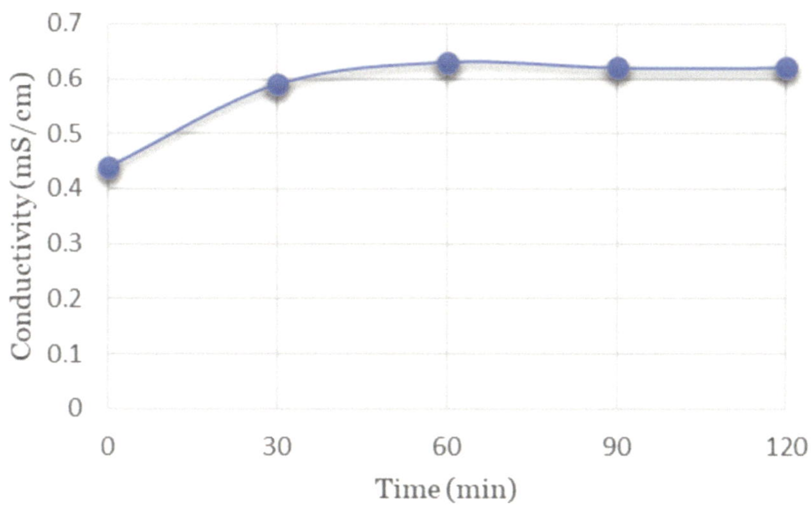

Figure 8.
Conductivity against time for experiments with trimethoprim.

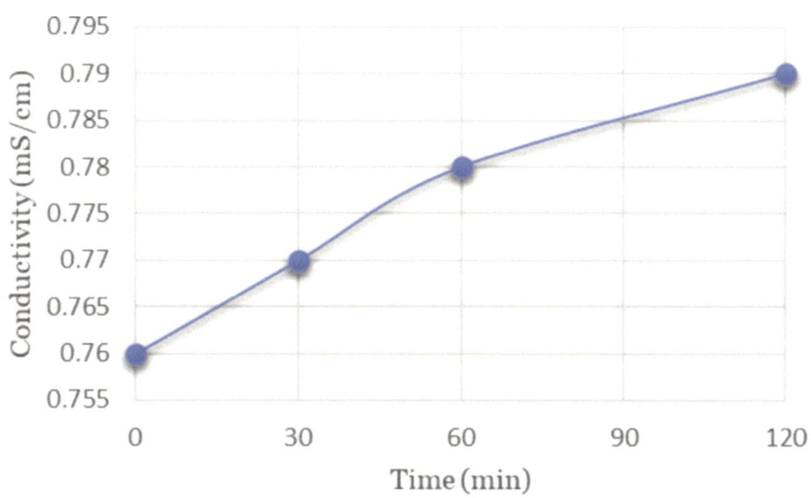

Figure 9.
Conductivity against time for experiments with the clavulanic acid-amoxicillin mixture.

199.1 mg/L and the second experiment (trimethoprim 0.1%) had the lowest drug concentration of 90.3 mg/L of COD; you can see that in 1 hour, the COD concentration of the first trimethoprim 0.2% experiment dropped to 142.8 mg/L, and in the second experiment, it dropped to a concentration of 88.5 mg/L. This shows that the degree of contamination of the water by organic compounds (TSWW) decreases, that is, the organic matter (drug) decreases, and therefore there is a break in the contaminant.

It is also observed that in the second experiment (trimethoprim 0.1%), there is a slightly lower concentration of drug and that when passing through the electrolysis treatment it presents a greater removal efficiency, since the COD decreases more as time passes, compared to the trimethoprim 0.2%, which has a slight increase and decay (**Figure 10**).

The results of total organic carbon can be seen in **Figure 11**. It is observed in the graph that there is a tendency to decrease the concentration of carbon as time goes by.

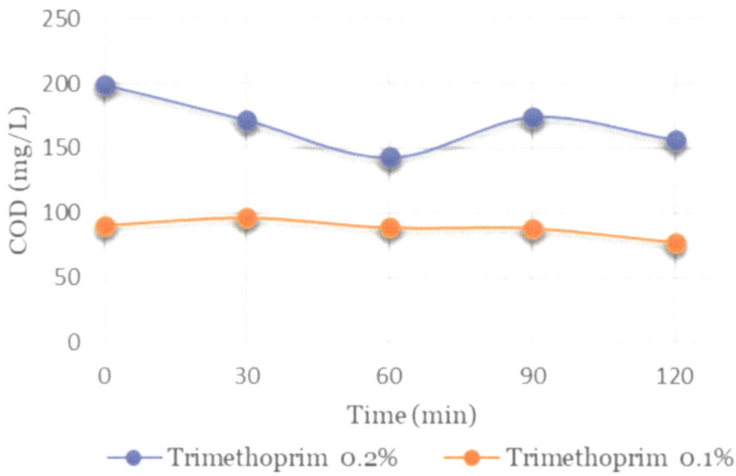

Figure 10.
Chemical oxygen demand with respect to time.

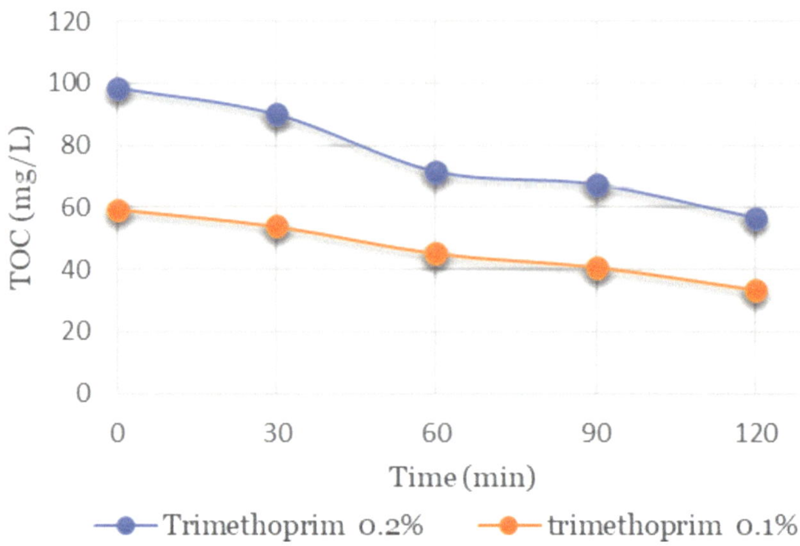

Figure 11.
Total organic carbon with respect to time.

TOC (mg/L)	
Trimethoprim 0.2%	**Trimethoprim 0.1%**
98.5	59.4
90	53.9
71.6	45.2
67.4	40.8
56.5	33.5

Table 1.
Total organic carbon results from experiments 1 and 2 with trimethoprim.

%Degradation according to TOC	
The first experiment (trimethoprim 0.2%)	**The second experiment (trimethoprim 0.1%)**
91.3706	90.7407

Table 2.
Maximum percentages of degradation obtained from TSWW samples with electrolysis treatment.

The initial TOC of the first experiment had a concentration of 98.5 mg/L and the second experiment had an initial concentration of 59.4 mg/L. In both cases, from the 30th minute on, it is observed that trimethoprim 0.2% decreases to a concentration of 56.5 mg/L and trimethoprim 0.1% to a concentration of 33.5 mg/L. The above proves that the electrolytic system is an alternative for trimethoprim decomposition.

Table 1 shows that during the electrolytic breakdown, carbon portions are mineralized but some fragments must have been ionized in different forms that affects the UV-Vis absorbance.

Finally, **Table 2** shows the maximum percentage of degradation that was obtained from the synthetic wastewater when it was treated with the electrolytic cell, for the first and second experiments with trimethoprim.

In the case of the clavulanic acid-amoxicillin TSWW mixture, the COD had an oscillating behavior because at 1 minute it began to decrease and at 60 minutes it abruptly increased and then it decreased again. The initial COD at time 0 for the clavulanic acid-amoxicillin TSWW sample (0.2%) has a concentration of

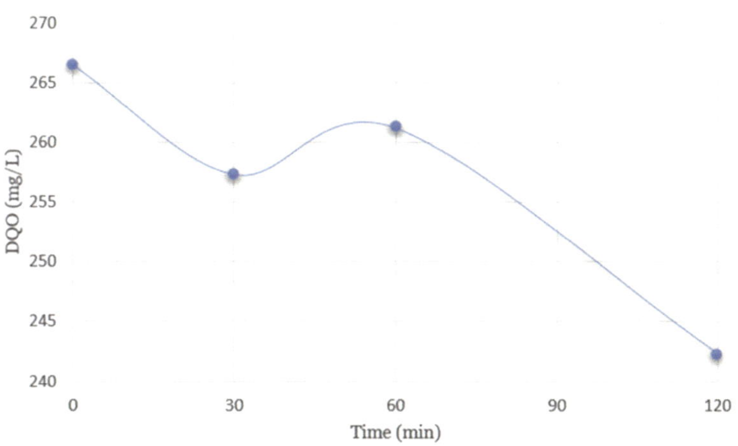

Figure 12.
Chemical oxygen demand against to time.

266.5 mg/L, after 30 minutes, there is a concentration of 257.3 mg/L, but it is observed that at 60 minutes, the COD concentration abruptly increased to 261.2 mg/L. This experiment ended his treatment with a concentration of 242.3 mg/L. According to the bibliography analyzed, the phenomenon obtained in **Figure 12** by the clavulanic acid-amoxicillin (TSWW) mixture could be due to the fact that when the sample is subjected to radiation, the molecule fragments and groups with free electron pairs are exposed and they absorb more radiation, generating increase in the values obtained. These formed fragments absorb a greater amount of energy after 60 minutes and as they continue to degrade and lose this capability.

The total organic carbon (TOC) results for clavulanic acid-amoxicillin TSWW can be seen in **Figure 13**; there is a tendency to decrease the carbon concentration as time passes. The initial TOC of the clavulanic acid-amoxicillin TSWW mixture had a concentration of 182.8 mg/L and drops to a concentration of 164.6 mg/L. The above indicates the decrease in the sample of organic compounds due to the breakdown of the organic molecules to molecules with simpler structures.

The results from the clavulanic acid-amoxicillin mixture experiments presented for the UV-Vis spectrophotometric studies give negative absorbance values, which is observed in **Figure 6**, probably due to the effects of the breakage of the molecule by chemical changes suffered in its structure, which generated ions or fragments with higher absorbances. This is because UV-Vis spectrophotometry excites the free electrons of oxygen, nitrogen, and sulfur that are present in both molecules, just as the molecule contains double bonds in its structure. In the pi bond, the electrons are excited and migrate to a higher energy level, modifying the structure. Therefore, the TOC was not used because negative values were being given in the absorbance.

The results of the clavulanic acid-amoxicillin mixture experiments presented for UV-Vis spectrophotometric studies in **Figure 6** give high absorbance values, probably due to the effects of the breakdown of the molecule due to the electrochemical changes suffered in its structure, whereby ions or fragments with higher absorbances are generated. The modification of its structure is due to the fact that UV-Vis spectrophotometry excites the oxygen, nitrogen, and sulfur free electrons that are present in both molecules, as well as the pi electrons of the conjugated double bonds in its structure. In the pi bond, the electrons are excited but they can be delocalized by resonance effects, modifying the energy absorbed by the structure.

During the reaction, the formation of free radicals like $HO^{\bullet}$ could occur, which are species with high reactivity, which allows them to attack organic molecules. It is

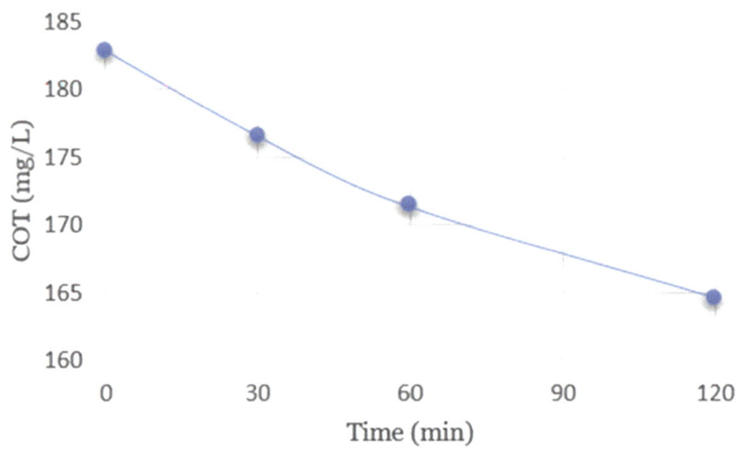

Figure 13.
Total organic carbon over time.

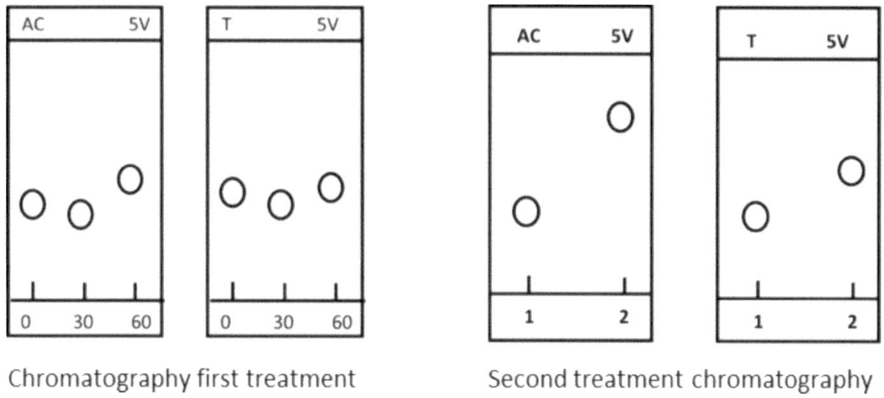

Figure 14.
Results of the thin layer chromatography demonstrating the rupture or breakdown of trimethoprim as also in the clavulanic acid-amoxicillin mixture, in which a great modification is observed, being less polar the product obtained with the treatment.

worth mentioning that the photochemical process is not developing in the electrolytic cell, despite the fact that TiO_2 reacts with light.

For the thin layer chromatography (see **Figure 14**), it is observed that there is a change of displacement between the 0 time and 60 minutes of the experiment. That is, the RF pattern for both drugs is not observed in comparison with the standards of trimethoprim and clavulanic acid-amoxicillin mixture; this indicates that the molecules at 60 minutes do not present the same structure. This is an expected result because in an electrolytic cell when an electric potential is applied, it generates an electric current passage, which is a flow of electrons between the electrodes creating a circuit of ionic and electric charges transport, besides getting the help of the electrolyte that maintains a balance of charges, which improves the transport of the charges between electrodes, causing the oxidation of fragments of the molecule in the surface of the anode and a reaction of reduction of other fragments of the molecule in the cathode.

In addition, the electrode that is doped with titanium dioxide generates a greater oxidative effect, debt to its catalytic properties and the mineral carbon, has a porous structure and a large contact surface, and allows an adsorption process to take place between the organic compounds and the mineral carbon. Another advantage of using charcoal is that the problem of forming products that may be toxic does not arise, due to its high absorption capacity as mentioned before, although its efficiency may depend on the amount of material organic present in the solution.

The electrolyte cell demonstrated based on the results that it has the feasibility of breaking or/and debugging emerging endocrine disrupting contaminants like trimethoprim. The work developed by Sirés et al. in 2005 on the electrochemical degradation of water acetaminophen by catalytic action of Fe^{2+}, Cu^{2+}, and light showed that the acidic aqueous solutions of the drug acetaminophen were degraded by anodic oxidation in an undivided electrolytic cell with a Pt anode and an O_2 supply [22], proving that the electrolysis process is functional. However, the materials and reagents to be used in this project are cheaper, and for a process tested in a single drug, there is no certainty how it will work in others.

Hirose et al. applied electrolysis for degradation of epirubicin, bleomycin, and mitomycin C with Pt/Ir electrodes and a NaCl solution as electrolyte. The results obtained were a partial degradation of the antibiotics but their cytotoxic and mutagenic activity was absolutely eliminated [23].

This research establishes that an elimination or rupture of different drugs can be carried out with an electrolytic cell, based on the work performed by Giraldo

Aguirre et al. in 2016 on the electrochemical treatment of waters containing β-lactam antibiotics, where by means of electrolysis, the degradation of the drugs oxacillin (OXA), cloxacillin (CLX), and dicloxacillin (DCX) was achieved. Those drugs are also used as antibiotics [7] like trimethoprim and the clavulanic acid-amoxicillin mixture and can affect aquatic systems and health due to the fact that antibiotics are among the most consumed and released drugs to the environment [24]. This release is also due to the unwarranted prescription with antibiotics, its inadequate dispensing, and the laxity in the regulation on the sale of medicines that allows self-medication with antibiotics, which are some of the factors that have been related to this high consumption [24] because it is a chemically defined substance capable of modifying the biochemical and physiological activity in an organism and therefore can produce a biological, beneficial, or toxic effect depending on the dose delivered [25]. Another problem is based on worldwide studies that have evidenced the presence of pharmaceutical compounds in effluents from fully operational treatment plants and with their respective control parameters, apparently operating with a high wastewater purification efficiency [26]; so, it is now known that with conventional wastewater treatments, it is not possible to eliminate this type of compounds in an efficient level [26], having as a consequence the proliferation of bacteria resistant to antibiotics, which will cause major problems in aspects of public health. In the research of Giraldo Aguirre et al. [7], a Ti/IrO_2 electrode was used which led to the oxidation process being better, which is due to the characteristics already mentioned for titanium and oxygen; also, the electrolyte that used sodium chloride helped them in the inhibition of microbial activity. This is because during the electrolysis reaction, chlorine gas is released from the electrolyte, but in our case, potassium hydroxide was used with the intention that the system can be incorporated to a wastewater treatment plant, since this technology usually uses biological reactors; therefore, only the breakdown of the molecule was sought without affecting the bio-catalytic bacteria during the subsequent processes. However, Giraldo Aguirre et al. conclude that electrochemical oxidation induces structural changes in antibiotic molecules and their results also indicate that electrochemical treatment is an effective technique for reducing the antibiotic potential that these compounds present, reducing the environmental risk due to the proliferation of bacteria resistant to antibiotics [7]. The aforementioned was also demonstrated within our experimentation when observed the graph of total organic carbon for the two experiments with trimethoprim and the thin layer chromatography for results with the clavulanic acid-amoxicillin mixture. This helps us show that the systems that use electrolysis break these compounds down into compounds that are less harmful to organisms and the environment. This experiment ends up being an important contribution to science in drug treatment and for a later use of electrolytic cells at higher levels, in addition to being highly versatile and to some extent economical oxidation processes.

4. Conclusion

The electrolytic cell modifies the structures of both molecules through different mechanisms, but in functional aspects, the electrolytic cell achieves the efficient degradation of trimethoprim unlike the clavulanic acid-amoxicillin mixture. It is established that the electrolytic cell can degrade some molecules more easily than others; so, we can establish that those compounds with a structure similar to trimethoprim could be degraded efficiently. However, it is also shown that to achieve the degradation of other compounds, it is necessary that the conditions of the electrolytic cell are adjusted again.

Addressing the problem of emerging pollutants that are endocrine disruptors is something really important and urgent, due to the impact they have not only on the environment but also on the health of living beings; so, it is important to transform them into less dangerous compounds for environment. Unfortunately, the methods of removing these contaminants are now expensive. That is why the implementation of the electrolytic cell is intended to create an economical option for the possible removal of these compounds, as it was shown to have a good degradation rate of trimethoprim. However, in the case of the clavulanic acid-amoxicillin mixture, it is necessary to modify its operating conditions.

Acknowledgements

All the authors appreciate the financial support given by the Instituto Politécnico Nacional (IPN) through the projects: SIP 20181685, SIP 20180081, and SIP 20190101.

Conflict of interest

The authors declare no conflict of interest.

Abbreviations

nm	nanometer
ml	milliliter
μm	micrometric
L	liter
g	gram
min	minute
Abs	absorbance
ECED	emerging contaminants endocrine disruptors
TOC	total organic carbon
COD	chemical oxygen demand
Redox	reduction-oxidation reaction
V	volts
SWW	untreated synthetic waste water
STWW	synthetic wastewater with treatment
TLC	thin-layer chromatography
mS	millisiemens

Author details

Jorge Alberto Mendoza Pérez[1*], Abril Gardenia Martínez Castillo[1], Jorge Octaviano Gomez Castrejon[2] and Juan Carlos Gómez Buendía[3]

1 Department of Environmental Systems Engineering, Escuela Nacional De Ciencias Biológicas From Instituto Politécnico Nacional, Mexico City, Mexico

2 Escuela Nacional De Ciencias Biológicas From Instituto Politécnico Nacional, Mexico City, Mexico

3 Department of Organic Chemistry, Escuela Nacional De Ciencias Biológicas From Instituto Politécnico Nacional, Mexico City, Mexico

*Address all correspondence to: jorgemendozaperez@yahoo.com

References

[1] Abdul-Mottaleb M, Mohammed JM, Abdul-Matin M, Musavvir-Arafat M, Mohammad AW. Emerging micro pollutant-pharmaceuticals and personal care products (PPCPs) contamination concerns in aquatic organisms—LC/MS and GC/MS analysis. In: Emerging Micro-Pollutants in the Environment: Occurrence, Fate, and Distribution. American Chemical Society Symposium Series, EE. UU. 2015. pp 43-74. Chapter 3. DOI: 10.1021/bk-2015-1198.ch003

[2] Muñoz-Macarena MFJ, Zahara MP, Alvarez-Torrellas S, Casas JA, Rodriguez JJ. Application of CWPO (catalytic wet peroxide oxidation) to the treatment of pharmaceutical emerging pollutants in different water matrices with a ferromagnetic catalyst. Journal of Hazardous Materials. 2017;**331**:45-54. DOI: 10.1016/j.jhazmat.2017.02.017

[3] Montes-Grajales D, Bernardes JL, Olivero-Verbel J. Urban endocrine disruptors targeting breast cancer proteins. Chemical Research in Toxicology. 2015;**29**(2):150-161. DOI: 10.1021/acs.chemrestox.5b00342

[4] Vademécum International Medicom. 14ª ed. Madrid: Medimedia-Medicom; 2014. EAN: 9788461736454

[5] Da Silva TH, Ribeiro AO, Nassar EJ, Trujillano R, Rives V, Vicente MA, et al. Nanocomposites of tetracyboximetalophalcyanin caoline/ TiO_2/cobalt (II) as heterogeneous photocatalysts for the breakdown of organic contaminants trimetoprim, caffeine and prometrin. Journal of the Brazilian Chemical Society. 2019;**30**: 2610-2623. DOI: 10.21577/ 0103-5053.20190178

[6] Tancawan AL, Pato MN, Abidin KZ, Asari AS, Thong TX, Kochhar P, et al. Amoxicillin/clavulanic acid for the treatment of odontogenic infections: A randomised study comparing efficacy and tolerability versus clindamycin. International Journal of Dentistry. 2015; **2015**:9. DOI: 10.1155/2015/472470

[7] Giraldo-Aguirre AL, Erazo-Erazo ED, Flórez-Acosta OA, Serna-Galvis EA, Torres Palma RA. Electrochemical treatment of water polluted with β-lactam antibiotics. Developmental Science. 2016;**7**(1):21-29. DOI: 10.19053/ 01217488.4227

[8] Barrera-Díaz C, Cañizares P, Fernández FJ, Natividad R, Rodrigo MA. Electrochemical advanced oxidation processes: An overview of the current applications to actual industrial effluents. Journal of the Mexican Chemical Society. 2014;**58**(3):256-275

[9] Fagan R, McCormack DE, Dionysiou DD, Pillai SC. A review of solar and visible light active TiO_2 photocatalysis for treating bacteria, cyanotoxins and contaminants of emerging concern. Materials Science in Semiconductor Processing. 2016;**42**:1. DOI: 10.1016/j.mssp.2015.07.052

[10] Balyejjusa S, Adome RO, Musoke D. Spectrophotometric determination of sulfamethoxazole and trimethoprim (cotrimoxazole) in binary mixtures and tablets. African Health Sciences. 2002; **2**(2):56-62

[11] Muchlisyam ESB, Raesa-Dathita R, Richa SP. Simultaneous assays of sulfamethoxazole and trimethoprim in suspension dosage form by three methods analytical of UV spectrophotometry. Rasayan Journal of Chemistry. 2019;**12**:1676-2357. DOI: 10.31788/RJC.2019.1245428

[12] Stojkovic G, Dimitrieska-Stojkovic E, Soklevska M, Velev R. Optimization, validation and application of UV-Vis spectrophotometric-colorimetric methods for determination of trimethoprim in different medicinal

products. Macedonian Veterinary Review. 2016;**39**:65-76

[13] Al-Degs YS, El-Sheikh AH, Harb DM. Accurate quantification of amoxicillin in different drug formulations using advanced chemometric methods. Jordan Journal of Pharmaceutical Sciences. 2020;**13**(1):21

[14] Gómez-Biedma S, Soria E, Vivó M. Electrochemistry analysis. Revista de Diagnóstico Biológico. 2002;**51**(1):18-27. ISSN: 0034-7973. Available from: http://scielo.isciii.es/scielo.php?script=sci_arttext&pid=S0034-79732002000100005&lng=es [Accessed: 14 March 2020]

[15] Aristizábal Fúquene A. Platinum: Scientific and socio-historic contributions XIX and XX centuries. Second part. Education in Chemistry. 2015;**26**(3):233-241. DOI: 10.1016/j.eq.2015.05.008. Available from: http://www.scielo.org.mx/scielo.php?script=sci_arttext&pid=S0187-893X2015000300233&lng=es [Accessed: 14 March 2020]

[16] Jaramaillo JD, Sánchez LEL, Amaris HV. Crystal structure of copper, mechanical and processing microscopic properties. Neogranadine Science and Engineering. 2006;**16**(2):96-103. Available from: https://www.redalyc.org/article.oa?id=91116210

[17] Centro Nacional de Metrología (CENAM) y entidad mexicana de acreditación a.c. (ema). Guía Técnica sobre Trazabilidad e Incertidumbre en las Mediciones Analíticas que emplea la técnica de Espectrofotometría de Ultravioleta-Visible. México, revisión 01; 2008

[18] Engels F. Dialéctica de la naturaleza [Internet]. Greenbooks Editore; 2019. Available from: https://books.google.com.mx/boks?id=gPC_DwAAQBAJ&pg=PT202&lpg=PT202&dq=formaci%C3%B3n+de+ozono+segun+Friedrich+Engels&source=bl&ots=lBV-sR_rbl&sig=ACfU3U3KOCUUT8exTCUE2F-qlSGvd3_QmQ&hl=es-419&sa=X&ved=2ahUKEwjx7ajm3arqAhUG16wKHfwCB1AQ6AEwAHoECA0QAQ#v=onepage&q=desintegraci%C3%B3n&f=true

[19] De la Cruz Solano HM, Valdez Vilchez M, Angulo Gutierrez O. Degradacion del azul de metileno, del efluente de la industria textile grupo grande sac, mediante oxidacion combinada ozono/H_2O_2 [Tesis]. de la Uviversidad Nacional del Centro de Perú; 2016

[20] Rao CN. Espectroscopia Ultravioleta y Visible. Vol. 2. Madrid: Alhambra, S. A.; 1970. pp. 19-28. ISBN: 9788420501987

[21] Millán F, Mérida I. CONCEPTOS Y PROCEDIMIENTOS DEL ANÁLISIS QUÍMICO CONTEMPORÁNEO III Evaluación de la espectrofotometría molecular UV-Vis. CITEIN Revista de Ciencia, Tecnología e Innovación. 2012;**5**:111-136

[22] Sirés I, Garrido JA, Cabot L, Centellas F. Electrochemical degradation of paracetamol from water by catalytic action of Fe^{2+}, Cu^{2+}, and UVA light on electrogenerated hydrogen peroxide. Journal of the Electrochemical Society. 2006;**153**(1):D1-D9. DOI: 10.1149/1.2130568J

[23] Hirose J, Kondo F, Nakano T, Kobayashi T, Hiro N, Ando Y, et al. Inactivation of antineoplastics in clinical wastewater by electrolysis. Chemosphere. 2005;**60**(8):1018-1024. DOI: 10.1016/j.chemosphere.2005.01.024

[24] Anahí D, Wirtz Veronika J, Corbett Kitty K, Gabriela E. Antibiotic use in Mexico: Review of problems and policies. Public Health Mexico. 2008;**50** (Suppl 4):S480-S487. Available from: http://www.scielo.org.mx/scielo.php?script=sci_arttext&pid=S0036-3634200

8001000009&lng=es [Accessed: 14 March 2020]

[25] Urbano G. Basic Principles of Pharmacology: Pharmacokinetics. 1st ed. San Cristóbal, Venezuela: Editorial Lito-Formas; 2011

[26] Angelina C, Lily M. Presence and elimination of pharmaceutical compounds in wastewater treatment plants: Global review and national perspective. Malariology and Environmental Health Bulletin. 2015; **55**(1):1-18. ISSN: 1690-4648. Available from: http://ve.scielo.org/scielo.php?script=sci_arttext&pid=S1690-46482015000100001&lng=es [Accessed: 14 March 2020]

Chapter 4

A Reverse Osmosis and Electrodialysis System Simultaneously Powered by Gravitational Potential Energy

Juvenal Rocha Dias
and Eliane Aparecida Faria Amaral Fadigas

Abstract

This chapter proposes an alternative system for conventional reverse osmosis (RO) and electrodialysis (ED) desalination plants by incorporating the use of gravitational potential energy (GPE). The proposed system is devised with two subsystems, the RO module followed by the ED module, both simultaneously powered by GPE. This kind of energy is obtained by storing the brackish water to be desalinated. The system's primary source of energy is wind. Windmills harness the wind energy to pump water to a reservoir located at a certain height (<20 m). The stored water has the GPE that will make a special plunger pump work. The piston of this special plunger pump is designed so that high pressure (about 15 bar) can be achieved in a different way from conventional RO plants. In the alternative system, here proposed, to pump water to the RO membranes, the special pistons go downward due to their own weight and are lifted, through a system of pulleys, with a counterweight filled with water obtained from the reservoir. The technical viability of the alternatives was theoretically proven by deductions based on physics and mathematics and with a special plunger pump prototype that worked successfully.

Keywords: wind energy, gravitational potential energy, reverse osmosis, electrodialysis, water desalination

1. Introduction

Desalination is a process through which pure water is obtained from water with a high concentration of dissolved salt (seawater or brackish water). In general, in a desalination plant, there is an inlet flux of salted water and two outlet fluxes (**Figure 1**).

One of the outlet fluxes (the brine) has a concentration higher than the inlet flux, and the other has a much lower concentration than the inlet flux (drinking water). To achieve that high reduction of solute concentration, it is necessary to deliver a certain amount of energy. The consumption of energy of a desalination plant is generally evaluated in terms of the quantity of energy used for each cubic meter of drinking water obtained. This parameter is called the specific energy consumption expressed

IntechOpen

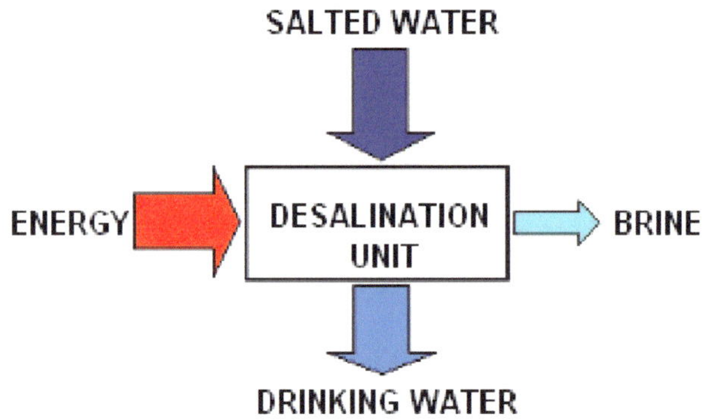

Figure 1.
Simplified scheme of a desalination unit.

in kWh/m^3. Among all desalination processes, we will briefly describe the reverse osmosis process for it is the one used with the system proposed in this chapter.

2. Reverse osmosis (RO)

Reverse osmosis is the reversal of a process that occurs naturally named osmosis. Osmosis occurs when two solutions with different concentrations are separated by a semipermeable membrane (**Figure 2**).

In such circumstances, there is a natural tendency to achieve the chemical equilibrium. Thus, the solvent crosses the membrane from the side of less concentration to the other side until the concentrations on both sides are the same. To reverse this natural tendency, a pressure (higher than the osmotic pressure) must be exerted on the side originally with higher concentration. This is what is done to achieve

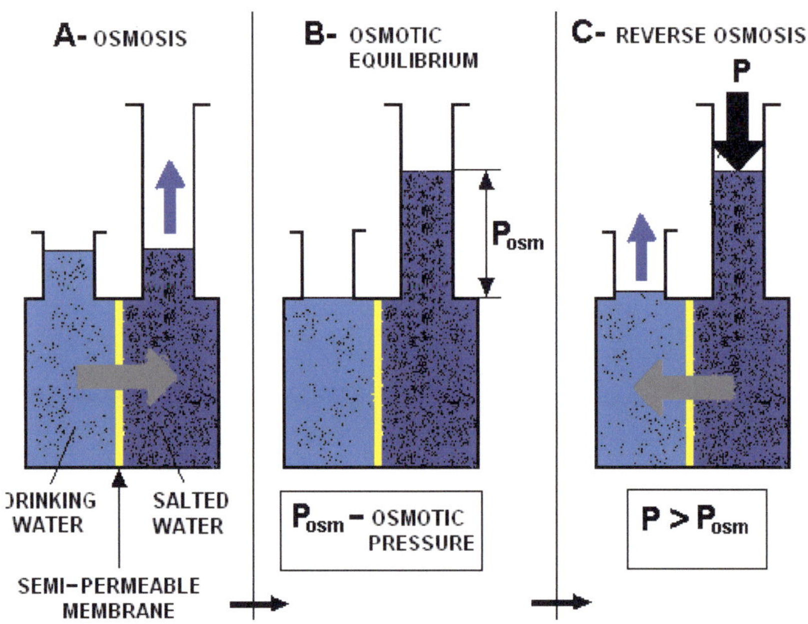

Figure 2.
Fluxes by osmosis and by reverse osmosis.

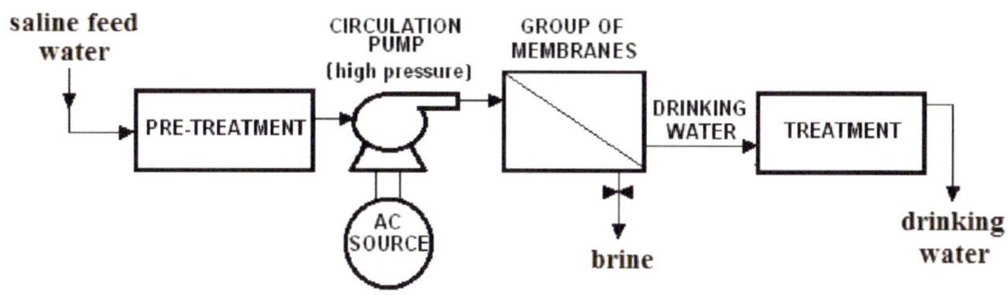

Figure 3.
Scheme of a reverse osmosis desalination plant.

desalination by reverse osmosis and why, generally, there is a high specific energy consumption. **Figure 3** shows a schematic of a RO desalination plant.

In this chapter, an alternative RO system powered by wind and gravitational potential energy (GPE) is proposed. The proposed system would basically replace the high-pressure circulation pump shown in the configuration of **Figure 3**.

3. Electrodialysis

The scientific principle that sustains electrodialysis (ED) is the electrical attractions between ions and electrodes of opposite electric charge. This phenomenon suits desalination since the dissolved particles in salted water are electrically charged (positive and negative ions). In ED desalination selective membranes are used to "sift" the ions. Some retain the positive ions and allow the negative ones to pass through, while others perform with the inverse characteristic. Membranes with inverse characteristic are placed alternately between the positive and negative electrodes. This configuration yields two separate outlet fluxes, drinking water, and brine. **Figure 4** depicts a conventional ED desalination module. ED is mostly used to desalinate brackish water.

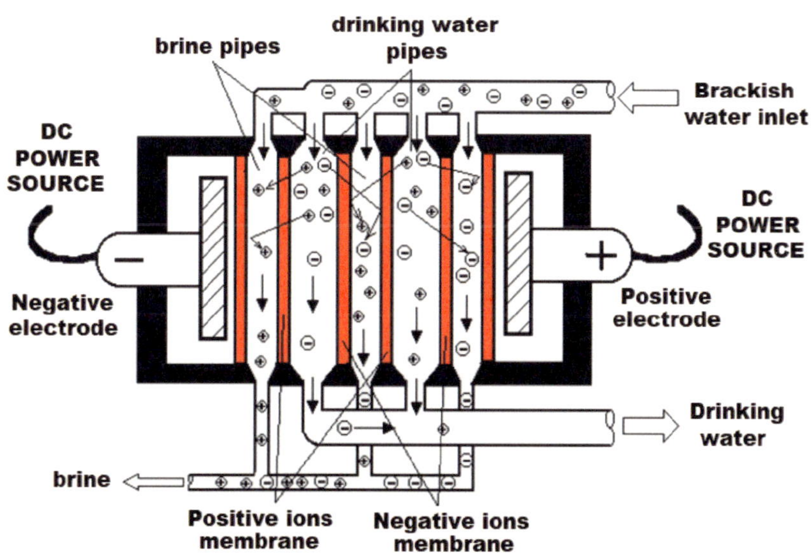

Figure 4.
Electrodialysis membrane configuration. Source [1].

4. The proposed system

Figure 5 shows the general configuration of the proposed system which simultaneously desalinates brackish water by RO and ED. The system's primary source of energy is wind power which is harnessed by windmills that pump brackish water to a reservoir at a certain height (less than 20 m). A subsystem (the hydraulic power column (HPC)) uses the gravitational potential energy of the stored water to achieve the necessary pressure (about 15 bar for brackish water) to feed the RO module. Notice that the flux of brine (still at high pressure) exiting the RO module is used to activate the Pelton wheel which, in turn, produces electricity to run the ED module.

Figure 6 shows the block diagram of the model outlining the hydraulic power column (HPC) where the increase in pressure occurs (from less than 2 bar to over 15 bar). As is technically required for RO and ED, before saline water reaches the membrane modules, it should undergo a pretreatment process to avoid premature malfunction. Cautiously, this feature should be integrated in the system both for the RO and the ED modules.

For seawater desalination, one needs to have an operation pressure of around 55 bar, while for brackish water it is around 20 bar. With the proposed system, this magnitude of pressure is given by a special piston working inside the HPC.

Figure 7 shows the proposed configuration for the HPC. The model is composed of two HPCs which, separately, can be seen as individual special piston pump. The two individual pumps are connected through a hydraulic automation system in

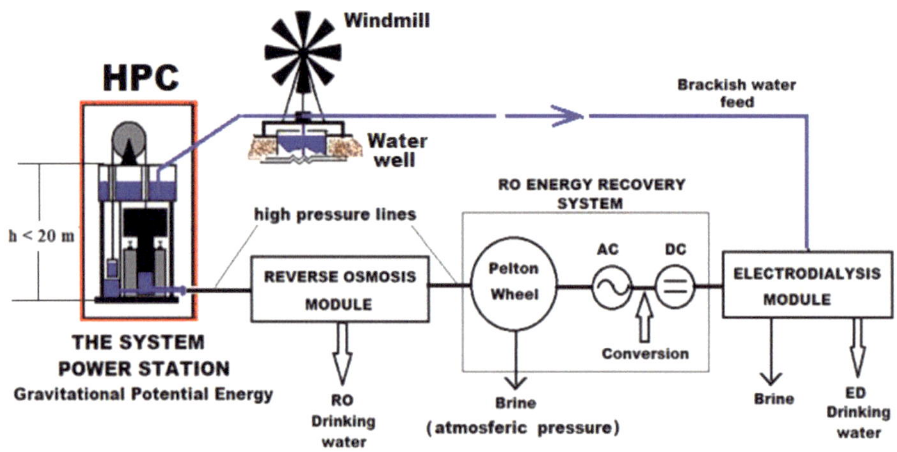

Figure 5.
The hydraulic power column renewable system.

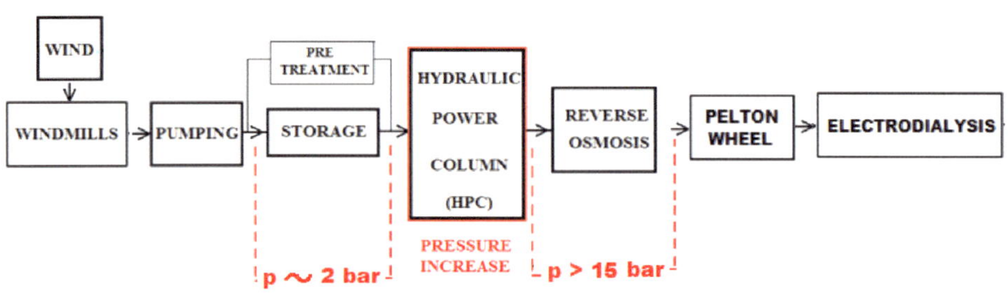

Figure 6.
The hydraulic power column renewable system.

DOI: http://dx.doi.org/10.5772/intechopen.91135

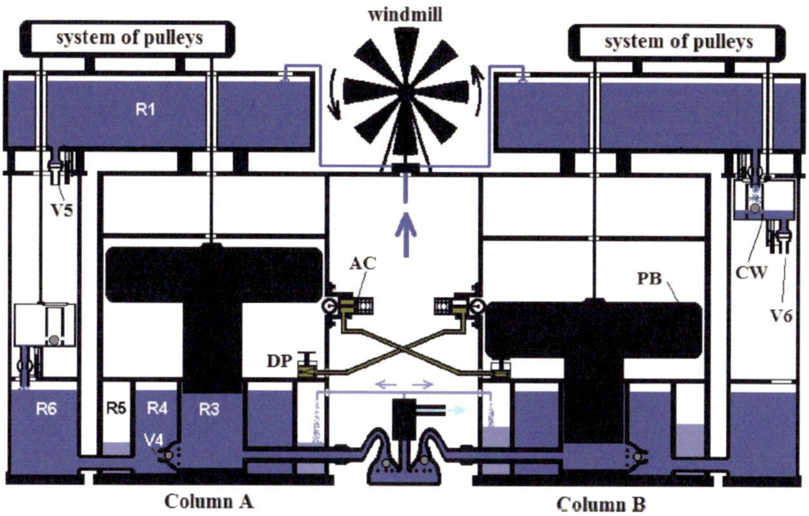

Figure 7.
Parallel operation of two hydraulic pressure columns. R1 fed with a windmill.

order to send a continuous flux of salted water to the modules of membrane where the desalination occurs.

The pretreatment of seawater, namely, filtering, is designed to occur by gravity in reservoir R1. After the filtering process, chemicals may be added if necessary. Note that if a reverse osmosis desalination system works with a low recovery rate (e.g., 20%), chemicals may not be needed. Recovery rate is the percentage of drinking water obtained from a volume of sea water that passes the membrane module (the rest is brine at high pressure).

With the system here proposed, to recover the brine energy, a Clark pump is used. It is a kind of flow work device. Brine at high pressure enters the Clark pump, leaves it at atmospheric pressure, and delivers work through a double piston pump. The system uses this work to pump water to the top reservoir R1.

Calculations have shown [2, 3] that it is possible to reach high pressures with a special piston of moderate dimensions. The counterweight gets the necessary volume of water (GPE) from the reservoir R1 to make the special piston go upward. The system of pulleys is designed so that it yields a reasonable force reduction to lift the piston. When the CW empties, the special piston goes downward, exerting pressure on the water in chamber R3. Consequently water is pumped into the reverse osmosis module under the designed and desired pressure, making reverse osmosis happen.

Shortly, during the parallel operation of columns A and B, (see **Figure** 7) there are the following events:

1. When the piston PB of column B reaches its bottom course, two events occur:

   - The counterweight CW is filled up because valve V5 opens.
   - The bottom of the piston rests on the dampers DP, pressing the fluid inside them. Simultaneously the actuators AC of column A retract due to the signal (fluid under pressure) received from DP, and the piston of that column is released.

Reciprocally, when the piston of column A reaches its bottom course, two identical events occur:

2. When the piston of column B, raised by the counterweight, reaches the upper course, two other events occur:

    - It is braked by the actuators and will remain in that position until, after filled up with water, it receives the signal sent by the actuators of column A.

    - Its CW empties because valve V6 opens.

An identical process takes place when the piston of column A comes to its upper course.

The hydraulic automatism guarantees the continuous function of the system as long as there is enough water in the reservoirs R1. The continuous flux of salted water to the modules of membrane is guaranteed by the two-way valve V7 that connects the two columns.

So far, a small prototype has been constructed at the polytechnic school of the *Universidade de São Paulo-USP in Brazil*. The prototype has proven that the mechanism of the HPC works. Since osmosis reverse is already a solid technique, the focus of the prototype was not to obtain drinking water because it would not be reasonable to reach 55 bar because of the size of the special piston.

Therefore the objective of the prototype was to prove that the HPC pumps water using the gravitational potential energy obtained from the water in R1 and that the necessary increase of pressure is achieved due to the shape of the special piston. The flux under the pressure exerted by the piston was used to drive a small Pelton wheel supported with magnetic levitation. This was done so, also to prove that the system can be used for electricity generation which is to be used to power the ED module.

5. The special pistons and their high pressure

Figure 8 shows the shape of the special piston that works inside the HPC. The HPC may work either with a massive special piston (piston B) or with one (piston A) where the increase of pressure is achieved mainly with the weight of water obtained from R1.

The pressure P_A exerted by the special piston A on its circular bottom of diameter ***de*** is a function of the geometric parameters illustrated in **Figure 8**. According to results already obtained [2], it is possible to achieve pressures high enough (over

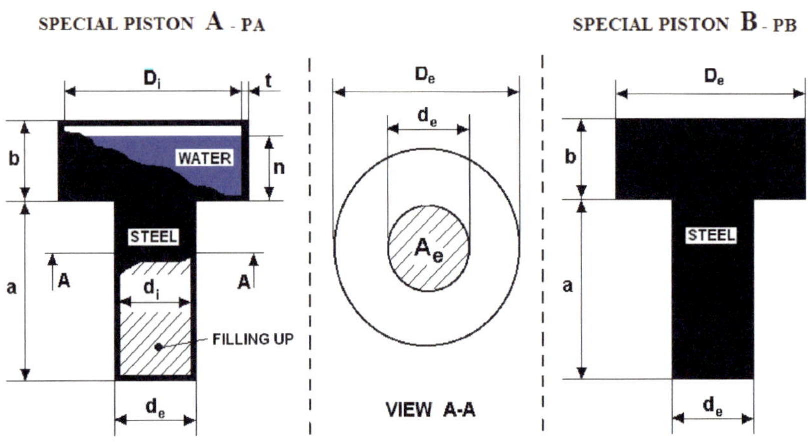

Figure 8.
Scheme of the special piston.

55 bar) to desalinate seawater with a special piston of moderate dimensions (fitting inside a 5-m-height HPC).

The expression for pressure P_A exerted by the special piston A on its bottom of diameter ***de*** is given by the total weight of the piston ***Pt***, divided by the circular area **Ae**:

$$P_A = \boldsymbol{Pt} / \mathbf{A_e} \tag{1}$$

Starting with Eq. (1) and knowing that ***Pt*** is the sum of the weight of steel, ***Psteel*** plus that of the water ***Pwater*** in R2 and the filling up material ***Pfill***, and that **Ae** = π/4 de^2, after some algebraic manipulation, one gets the following relation [2, 3]:

$$\mathbf{P_A} = \frac{g}{de^2} [\rho_{stl} \, (De^2 b + de^2 a)] - Di^2 (b - 2t) - di^2 a] + \rho_{wat} Di^2 n + \rho_{fill} \, di^2 a] \tag{2}$$

where **g** is the acceleration of gravity [m/s²]; **ρstl** is the density of steel [kg/m³]; **ρwat** is the density of seawater (or brackish water) [kg/m³]; **ρfill** is the density of the of the filling up material [kg/m³]; and **Di – De -2 t** [m], being **t** the thickness of the piston's wall.

A similar approach with special piston B leads to the following expression for P_B:

$$\mathbf{P_B} = \frac{g}{de^2} \rho_{stl} \, (De^2 b + de^2 a) \tag{3}$$

A very interesting aspect of using special piston A is that the filling up material may be anything available where the system will operate such as stones, sand, or even recycled metals. This low-cost characteristic is useful for remote areas because it would not be necessary to transport a very heavy structure.

6. The system's specific energy consumption regarding the RO module

The analysis of a configuration for an underground HPC (**Figure 9**) led us to some conclusions about the specific consumption of energy [2, 3]. Notice that there is no need to pump water to R1, and the energy consumption is used only to lift the piston.

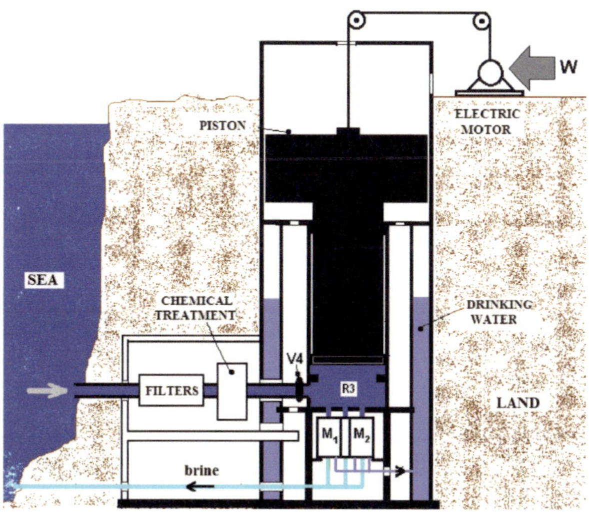

Figure 9.
Underground HPC.

- Considering a frictionless piston with an operating pressure of 20.3 bar, the system would consume around 0.563 kWh for each cubic meter of drinking water which is the minimum allowed by nature [4].

- Considering a frictionless piston with an operating pressure of 55 bar, the HPC would consume around 1528 kWh for each cubic meter of drinking water.

- For an operating pressure of 55 bar and taking into account acceptable efficiencies ("actual" values) for the components of the HPC, it has been shown that the predicted "actual" specific energy consumption is about 2811 kWh for each cubic meter of drinking water. This is a very good result compared to the conventional RO systems for which this parameter is generally above 4 kWh/m^3. Taking into account the ED module, we can predict an overall energy consumption about 3 kWh/ m^3 of drinking water.

7. Calculation of the systems' pumping power

Figure 10 shows the chain of energy transformations that occur in in the proposed system. Starting from wind energy until the storage of gravitational potential energy (GPE) in the reservoir R1 on top of the hydraulic power column.

In the block diagram of energy transformation, the conversion of wind power P_w into pumping power P_p that will be used to feed R1 with water to be pumped from a well is illustrated. Along those conversions, there are power losses due to the inefficiencies of the devices and mechanical connections (shafts, gears, etc.). Further in this chapter, it will be shown how the pumping power P_p is converted into a volume of water V_w, stored in R1.

The power of wind flow P_w is given by [5]:

$$P_w = 1/2\ \rho\ A\ V^3 \tag{4}$$

where **ρ** is the air density; **A** is the frontal area through which the air flows; and **V** is the wind speed.

A wind machine (windmill or wind turbine) can be used to harvest this wind power and convert it into its rotor power P_r. Betz has proven that 59% of P_w is the maximum P_r that an ideal rotor would be able to harvest from the wind [5]. Observing **Figures 10** and **11**, one concludes that for system 1 (windmill), P_r is used to drive a piston pump which will feed R1, while in system 2 (wind turbine), it will be converted into an electric power P_e that will power an electric pump which feeds R1.

Back to the rotor power, its magnitude is:

$$P_r = C_p\ 1/2\ \rho\ A\ V^3 \tag{5}$$

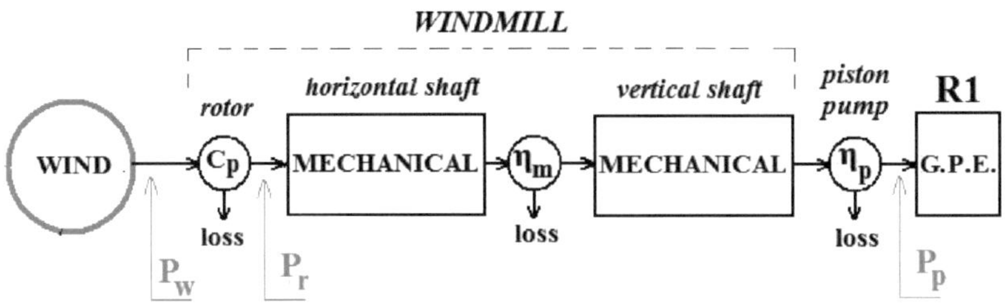

Figure 10.
Energy conversions of the system.

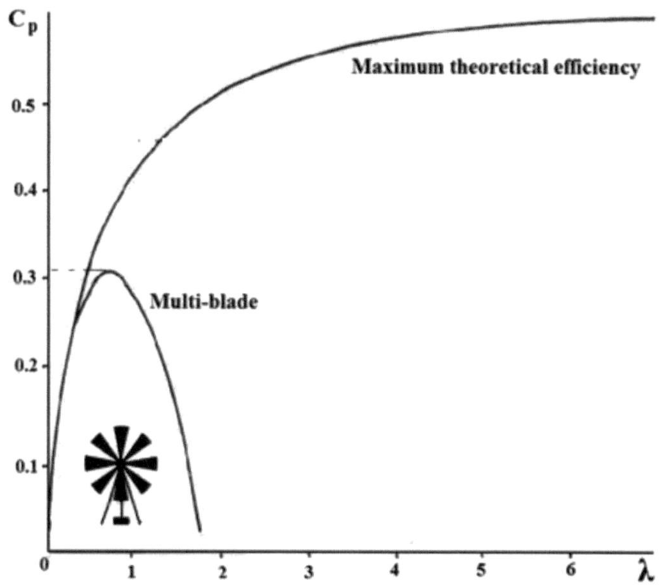

Figure 11.
Aerodynamic efficiency of wind machines: Source [5] with our drawings.

where C_p is the power coefficient of the machine. C_p can generally be expressed as a function of the tip speed ratio, λ, defined by [5]:

$$\lambda = \text{blade tip speed/wind speed} = \Omega R/V \tag{6}$$

where Ω is the angular velocity of the wind rotor and R is the radius of the wind rotor.

Figure 11 illustrates the aerodynamic efficiency of a windmill which is characterized by its power coefficient C_p. Resortint to **Figure 11**, a windmill's C_p can be compared with Betz's maximum theoretical efficiency.

According to **Figure 10**, taking into account the efficiencies of the gear η_m and that of the piston pump η_p, the pumping power P_p is given by [6]:

$$P_p = P_r . \eta_m . \eta_p \tag{7}$$

So, the pumping power of the system with windmill is:

$$P_p = C_p\, 1/2\, \rho\, A\, V^3\, \eta_m . \eta_p \tag{8}$$

Eq. (8) represents the theoretical pumping power P_p at the end of the chain of conversions shown in **Figure 10** for the system. However, in practice one can resort to other equations that are based on the characteristics of each wind machine used by the system. Since windmills are generally used to pump water, it is sufficient to determine its pumping power P_p. For S2, first it is necessary to determine the electric power P_e delivered by the wind turbine and then estimate the efficiency of its connection with the electric pump in order to find the pumping power.

8. From pumping power to potential gravitational energy stored in R1

After finding the pumping power P_p, one has to find the flux of pumped water to R1 as a function of the manometric height Hm (head) as well as the volume of

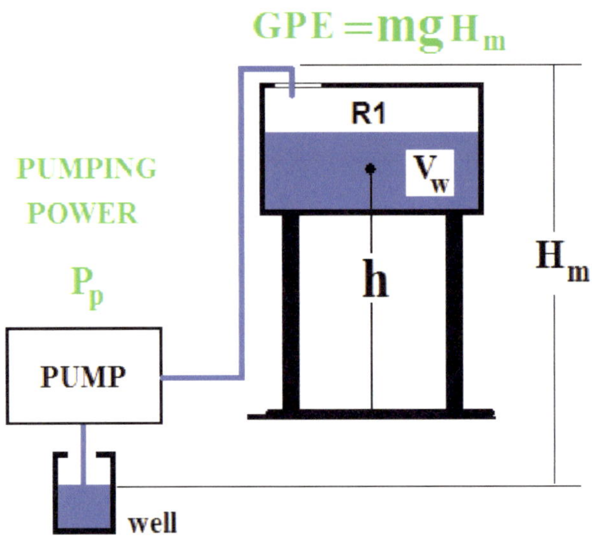

Figure 12.
Storage of gravitational potential energy in reservoir R1.

water V_w stored in R1 (**Figure 12**). During the pumping process, the work done by the pump is equivalent to energy transferred (E_{tra}) to the water which will be stored in R1 as gravitational potential energy (GPE).

The energy transferred E_{tra} to the water during the pumping process is:

$$E_{tra} = P_p . \Delta t. \tag{9}$$

where Δt is time (in seconds) corresponding to the period of operation of the pump [hour(s), day(s), month(s), year(s), etc.].

As mentioned, the transferred energy is stored as potential gravitational energy (GPE) in R1. Therefore one has:

$$E_{tra} = GPE, \tag{10}$$

or

$$P_p . \Delta t = m . g . H_m. \tag{11}$$

where H_m (the head) is the height that the pump must overcome in order to force water from the well to the top reservoir R1 (**Figure 12**).

Since mass of pumped water is the product of the specific mass for its volume ($m = \rho_w V$), Eq. (11) becomes:

$$P_p . \Delta t = \rho_w V_w g H_m. \tag{12}$$

Solving Eq. (12) to obtain the volume of a pumped water, one has:

$$V_w = P_p . \Delta t / (\rho_w . g . H_m) \tag{13}$$

where **Vw** is the volume of pumped water (stored in R1); **Pp** is the pumping power; **ρw** is the water density; **g** is the acceleration of gravity; and **Hm** is the manometric height.

If one considers the time of operation of 1 hour ($\Delta t = 3600$ s), Eq. (13) becomes:

$$V_w = 3600\ P_p./(\rho_a.g.H_m) \quad (14)$$

With $\rho_w = 1025$ kg/m^3 corresponding to the density of seawater to be pumped to R1, and.

$g = 9.81$, Eq. (14) reduces to:

$$V_w = P_p/(2790\ H_m) \quad (15)$$

Eq. (15) yields the volume of pumped water per hour. Therefore the result of this equation should be multiplied by 24 to obtain the volume of water stored in R1 during a day. Once there is enough water in R1, to make the hydraulic power column work, as already proven with the prototype, the reverse osmosis process will occur as long as the special piston exerts the necessary pressure. For that, the pressure is previously calculated according to the design of the piston and using Eqs. (2) and (3).

9. Alternatives to get the potential gravitational energy stored in R1

The HPC is designed so that it works as long as there is enough water stored in the reservoir R1 regardless of how water was put in it. This characteristic yields a certain versatility to the whole system. This means that one could feed R1 with other sources of energy such as wind, sea waves, sun, etc. In this regard, it is interesting to consider the possibility of using a device similar to the HPC's plunger pump to harness sea wave power to pump water to reservoir R1. Such device is currently a reality due to the invention of Alvin Smith who developed the Searaser (**Figure 13**). This feature is especially important if the system is to be designed for seawater desalination with the RO module and brackish water for the ED module.

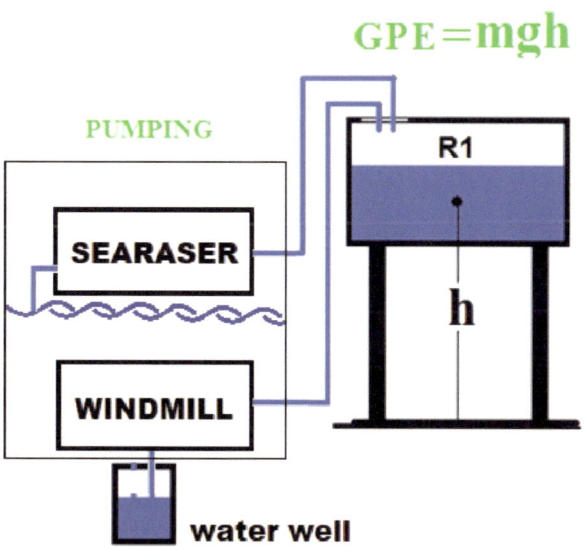

Figure 13.
Feeding of reservoir R1 with windmill and Searaser.

10. Conclusion

From what has been done so far, we conclude that the model is technically and economically practicable. The economic viability is supported, not only by the fact that the model uses a renewable "free" source of energy, but also because of a predicted low specific energy consumption (around 2.811 kWh/m^3) for the RO module, regardless of which system is used to feed the top reservoir R1. Taking into account the ED module, we can predict an overall energy consumption about 3 kWh/m^3 of drinking water.

Inherent to the use of wind gravitational potential energy is the reduction of emissions of CO_2 which is an environmental advantage compared to conventional RO plants. Another advantageous fact is a low cost of maintenance due to the simplicity of the hydraulic power column. The major drawback of the model is the need for a big area to install the windmills if one needs to desalinate a large quantity of water. This disadvantage can be overcome using a hybrid source of energy to feed R1 such as Wind-Sea-Sun combination. The constructed prototype has proven that the system is practicable.

Author details

Juvenal Rocha Dias and Eliane Aparecida Faria Amaral Fadigas*
Department of Electrical Automation and Energy Engineering, São Paulo University, São Paulo, Brazil

*Address all correspondence to: diasjuvenal@yahoo.com; eliane.fadigas@poli.usp.br

References

[1] Dias JR. Osmose Reversa com Energia Eolica e Energia da Gravidade. Germany: Novas Edições Acadêmicas; 2017

[2] Dias JR. Dessalinização de água por osmose reversa usando energia potencial gravitacional e energia eólica [Dissertação de Mestrado]. Brasil: Defendida na Universidade de São Paulo; 2004. p. 212

[3] Fadigas EAFA, Dias JR. Desalination of water by reverse osmosis using gravitational potential energy and wind energy. Desalination. 2009;**237**:140-146. Available from: www.sciencedirect.com

[4] Isentropic Desalination (At Minimal Energy Cost) of Seawater. Available from: http://mshades.free.fr/isentropiques/desalination.html [Accessed: 15 March 2004]

[5] Manwell JF, McGowan JG, Rogers AL. Wind Energy Explained. Amherst, USA: John Wiley & Sons Ltd., University of Massachusetts; 2002

[6] Fadigas EAFA, Dias JR. A nonconventional reverse osmosis plant powered by storage of wind energy as gravitational potential energy. In: Desalination: Methods, Costs and Technology. New York: Nova Science Publishers. Inc.; 2010. pp. 229-245

Chapter 5

Toxic Effluent Treatment by Membrane Based Ultrafiltration and Reverse Osmosis for Sustainable Management and Conservation of Ground Water in Industrial Clusters

Biman Gati Gupta

Abstract

The present study attempts to assess the nature of effluents generated from textile bleaching and dyeing units located at Kalikapur area under Maheshtala region, West Bengal, India and to provide a sustainable management of ground water resources through installing CETPs with zero liquid discharge system. Effluent from medium, small and tiny units of this region is estimated at 2000 MLD. Studies with 40 units for 4 years (2012—2016) located in this area exhibited following mean values of different physic-chemical variables: pH (9), Biological Oxygen Demand (610 Mg/L), Chemical Oxygen Demand (1827 Mg/L), Total Dissolved Solids (6411 Mg/L), Total Suspend Solids (927 Mg/L) and toxic metals such as lead Pb (0.43 Mg/L), Chromium (0.031 Mg/L), Zinc (0.74 Mg/L), Nickel (0.07 Mg/L) and Cadmium (0.03 Mg/L). These finding of results surpass the standard allowable limits qualify by FAO (1985) and World Health Organization (2003). The waste water loaded with toxic trace metals is adversely affecting the environmental pollution and anthropomorphic eudemonia and also pollute the quality of both surface and ground water and consequently degraded agricultural and plant yield, vegetable and fruits and causes impairment to aquatic lives. Four to five Common Effluent Treatment Plants are urgently required to install at different areas of the Maheshtala cluster with a capacity of 500 MLD each, so that one in Kalikapur area, to manage sizeable volume of waste water (2000 MLD) and sustainable management of ground water resources in a thickly populated urban area near Calcutta, a principal city of India.

Keywords: textile, toxic, effluent, water, contamination, treatment, membrane, ultrafiltration, reverse osmosis

1. Introduction

The main aim of the study is to assess the nature of waste water generated and to provide a realistic sustainable groundwater management by installing common effluent treatment plant with zero liquid discharge system through implementation of membrane based ultra-filtration, reverse osmosis with recourse to recycling of bleaching and dying effluent at Kalikapur, West Bengal, India to save groundwater, environment and human health of a thickly populated area and 15 KM from Calcutta, a premier city of India. It revealed during study that the contamination level of surface water including canal water running in the cluster is highly contaminated. The waste water discharged into the nearby canal water is the main source of contamination as most of the units do not treat the waste water due to lack of treatment facilities in their own. The soil profile of the area is also degraded in the locality. Hence, the only way left behind is to treat the waste water to a level which can be reused in textile units through recycling. Membrane based treatment of waste water is found suitable in this circumstances to treat the toxic waste water as well as recycling of the treated water to the textile units to save groundwater paucity of the area. The water requirement in these industries is very high compare to any other industries in the world. The area Maheshtala, Chatta, Kalikapur, Mahishgot lie in South 24 Parganas district of West Bengal (**Figures 1** and **2**) between 10.45^0 N latitude to 75.90^0 E longitude having more than 1400 small and tiny bleaching and dying units as per Economic Survey (2014), Govt. of West Bengal where groundwater table is shrinking day by day as per SWID, Govt. of West Bengal survey.

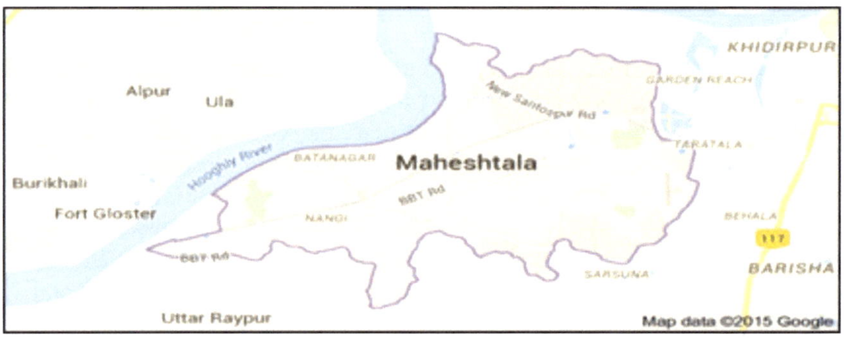

Figure 1.
Map of Maheshtala region.

Figure 2.
Map of Chatta Kalikapur.

2. Assessment of waste water quality

The overall mean concentrations of different physico-chemical parameters studied (pH, TDS, BOD, COD, NO3, hardness, Fe, F, Pb, Cd, Cr and Na) showed distinct seasonal trend in the following order of variation: pre-monsoon (April, May and June) > monsoon (July, August September and October) > post-monsoon (November, December, January and February) ($P < 0.05$). The physicochemical profile of waste water reflects that highest concentrations of both metals and organic loading are associated with the quantum and quality of bleaching and dyeing effluents released from the units studied. For the Maheshtala textile units the manufacturing processes like cutting and stitching activities of cloths start in the pre-monsoon season whereas the bleaching and dyeing processes starts in June continues till August stretching pre-monsoon to monsoon to meet the demand in festival season, which have significant implications in both metal and organic loading in the wastewater effluents. The present findings on high metal concentration and deteriorated physicochemical parameters are similar in other studies on textile industries waste water [1–3].

There appeared significant differences in metal concentrations in wastewater reflected in the following order: Na > Fe > Cr > Pb > Cd ($P < 0.05$). The metals in wastewater find their sources in the chemicals (like sodium hydrochlorides, sodium hydro-sulphites, optical brightening agent, caustic soda etc.) and dyes (mordant dyes, azo dyes, disperse dyes, vat dyes, indigo dyes etc.) used for the wet processing of textiles.

The heavy metal pollution index (HPI) for the wastewater studied is estimated as 926 which exceeds the critical level of 100 for drinking water as determined by Hakanson model [4] by about 9 folds. The study has shown similar HPI value for water of Jamuna river (near Delhi) obtained by [5, 6] following the same model (1492). Again the metal index (MI) in case of the studied wastewater is 4.68 which far exceeds than that of the threshold level of 1 for drinking water as prescribed by Hakanson model [4]. While assessing the water quality index (WQI) the results show that the wastewater poses severe potential ecological/health risk during all the seasons being maximum threat (WQI PRM-16) followed by (WQI M-19) and the minimum in post-monsoon (WQI PSM-31). According to Hakanson model (in the scale of 0–100) the present risk falls under the severe category in pre-monsoon and monsoon and critical in post monsoon with an annual average WQI being 22.

While comparing with the permissible limits for relevant parameters in case of wastewater discharge as per IS: 10500 (2012) it was observed that the values of the physico-chemical parameters and metals in the present study exceeded the respective limits.

All the physico-chemical parameters and metal concentrations in wastewater are very much higher than the permissible limits for drinking water of BS: 10500 (2012) and WHO (2003). These indicate that wastewater is critically contaminated compared to limits provided for drinking water. It finds conformity with observations of [7]. Due to higher level of TSS, TDS, BOD, COD and presence of very high concentrations of Pb, Fe and Na the wastewater becomes highly contaminated and often toxic, and there remains high chance of transfer of metals to soils and subsequently to crops, vegetables and fruits when irrigation contamination occurs due to advertent irrigation or inadvertent flooding of agricultural fields [8]. Wastewater also contaminates the water ways due to its direct discharge to the Chatta canal and damage aquatic ecosystem of canal, pond, watershed and ground water due to infiltration as shown by similar study conducted by [9].

Statistical correlation matrix shows that there is a significant positive correlation (r = 0.60) of Cd with TDS. Sodium has significant positive correlation with TSS (r = 0.78) and Fe (r = 0.56) and negative correlation with fluoride (r = −0.87).

3. Assessment of quality of canal water

The overall mean data of different physicochemical parameters of canal water indicate that TSS, TDS, BOD, COD, CaCO3, Na, Pb, Cd and Cr are higher than their permissible limits for drinking water prescribed by IS: 10500 (2012) and WHO (2003).

There appears a distinct seasonal trend in BOD, COD and metal concentrations in canal water in the following order of variation: PSM > M > PRM ($P < 0.05$). This finding can be explained by the fact that as the manufacturing as well as bleaching-dyeing activities become more intense during November and December to meet the demands of the Christmas and Ramjan festivals compared to the B and D activities in monsoon for meeting the demands of puja festive season the quantity of wastewater in the former becomes greater inclusive of its metal and organic load. Their concentrations fall during pre-monsoon due to reduced production and concomitant effluent discharge load.

There appeared significant differences in metal concentrations in canal water reflected in the following order: Na > Fe > Pb > Cr > Cd ($P < 0.05$). The metals in wastewater find their sources in the chemicals like sodium hydrochlorides, sodium hydro-sulphites, optical brightening agent, caustic soda etc. and dyes (mordant dyes, azo dyes, disperse dyes, vat dyes, indigo dyes etc.) used for the wet processing of textiles.

Further the overall mean values in mg/L of TSS (37), TDS (2150), BOD (60), COD (293), Na (669), metals (Pb, Cd and Cr) specify the canal water is very much contaminated similar to observations made by [10]. TDS in canal water appears due to presence of mostly dissolved inorganic salts (principally calcium, magnesium, potassium, sodium, bicarbonates, chlorides and sulfates) and partly contributed by small amounts of organic matter dissolved in water [11].

The heavy metal pollution index (HPI) estimated for the canal water is 689 which has far exceeded (about 7 fold) the critical pollution value of 100 as per Hakanson model [4]. The metal index (MI) for canal water has been calculated as 6.90 which is about 7 times the critical pollution value considering all the metals present as per Hakanson model [4]. The water quality index (WQI) for canal water indicates that there appears potential ecological/health risk during PRM (49), M (39) and PSM (28) indicating severe water quality in pre-monsoon and monsoon, and critically deteriorated condition of water quality during post-monsoon season as per Hakanson model [4].

The high values of COD and BOD are attributed to the organic loading and oxidizable matter present in wasterwater discharged from the B&D units coming to the canal water and its mixing with the sewage coming from neighboring region. The very low level of dissolved oxygen (1.7 mg/L) is indicative of oxygen deficiency in the canal water which appears to be detrimental for sustenance of biota excepting a few indicator species [12]. During monsoon season as the canal overflows the canal water contaminates adjoining agricultural fields, it contaminates them due to presence of metals like Pb, Cd, Cr, Fe and Na. Such build-up of contaminants has serious adverse implications for severe contamination of the fields being almost agriculturally nonproductive; even the vegetables and fruits grown in selected residents of that area were found heavily laden with metals. The presence of total coliforms and fecal coliform indicates that the canal water was microbiologically contaminated, which may have their genesis from contamination of B&D wastewater with domestic sewage and fecal matters. The canal water can pollute the ground water due to presence of metals (Pb, Cd, Cr, Fe and Na) and inorganic chemicals due to long term infiltration [13]. These findings are similar in other studies on contamination of surface water due to textile industries [14–21].

The statistical correlation analyses show that COD has significant positive correlation with BOD ($r = 0.98$). Na has significant strong positive correlation with BOD ($r = 0.91$) and COD ($r = 0.98$) while Cr has positive correlation with Na ($r = 0.47$). Again nitrate has significant positive correlation with BOD ($r = 0.91$), Na ($r = 0.73$) and negative correlation with pH ($r = -0.45$).

4. Assessment of quality of soil

The overall mean data of different metals and chemical parameters of soil indicate that Pb, Cr, Cd, Zn, Fe, NO_3 are higher than the permissible limit of WHO (1996) The heavy metal concentrations in soil show the pattern in the following order: Zn > Pb > Cr > Fe > Cd ($P < 0.05$). These indicate that soil is contaminated with metals and not suitable for agricultural production [22]. The metal index for soil has been estimated as 336, which far exceeds the critical pollution value of 100 as per Hakanson model [4]. The soil quality index (SQI) indicates that it poses a potential ecological/health risk as reflected from its value (519), which is about five time the value of critical condition i.e. >100 due to presence of toxic metals. Among the metals highest ecologically risk is posed due to Cr and Pb followed by moderate risk from Zn and slight risk due to Cd. Considering the relative heavy metal concentrations in the canal water and soil the heavy metal concentration factors for different metals show the highest concentration in case of Cr (2325) followed by Pb (314) and Cd (44.24). Such metal concentrations in soil in course of time get transferred to vegetables and fruits grown in the locality and pose severe health risks to humans [23]. These results can be corroborated with the findings of several other studies [24–27]. Statistical correlations show that Cr is positively correlated with nitrate (NO3) ($r = 0.57$) while and F is negatively correlated with NO3 ($r = -0.49$).

Bleaching and dyeing units consume major quantities of water in textile industries, is the third largest user of water in the world. Wet processing is one of the major one in textile engineering. In all stages of wet processing large amount of water is used. Soft water is the major requirement of wet processing in textile industries for leading production quality.

5. Some important parameters of water for textile processing

5.1 pH

pH is the reference point of H+ ions concentration, its quantity indicates the quality of water,such as neutral, acidic or alkaline. .pH of neutral at 7, less than 7 indicates acidic and alkaline when above 7. The pH measurement scale is having value from 0 to 14.

5.2 Turbidity

Turbidity is reason by the scattering of light by suspended substance which may be organic or inorganic in universe. The turbidity of water is calculated against a standard solution having a standard turbidity value 1000 units.

5.3 Color

Color normally shows the presence of suspended and soluble matter, which affects the wet processing. The Hazen unit is the measurement of color of water and is comparing it with a color of a standard water solution. A colored Hazen unit is produced by liquefaction 1 ppm platinum in the form of chloroplatinic acid, in the impression of 2 ppm cobalt chloride.

5.4 Total dissolved solids (TDS)

TDS contain of small amounts of organic matter and inorganic salts that are change state in water. The TDS is calculated in ppm (Mg/L).

5.5 Total suspended solids (TSS)

The suspended solids are distinct particles which are not soluble in water. These can be separated by filtration and are also calibrated in ppm.

5.6 Acidity

Major natural waters are buffered by a HCO_3/CO_2 system. Carbon acid is not amply neutralized until a pH of 8.2 and will not depress pH below 4.5. CO_2 acidity is in the pH range of 8.2–4.5. The reason of mineral acidity caused in industrial waste having pH is below 4.5.

5.7 Alkalinity

The alkalinity is due to the existence of bicarbonates (HCO_3), carbonates or hydroxides. Alkalinity is separated into caustic alkalinity (above pH 8.2) and total alkalinity above pH 4.5.

6. Disadvantages of hard water in textile wet processing

Shade change caused due to formation of hard soaps with calcium and magnesium ions. Carbonates of calcium and magnesium precipitates iron (Fe) and aluminum (Al) black and essential cotton dyestuffs. Hard water creates dyes duller and even form scum. The metals Fe and Cu ion are known impurities, is a problem in the peroxide bleaching, Fe (iron) is responsible for reducing the brightness level of many dyes and is also objectionable in the washing off activities in wet processing.

Hard water is accountable for scaling in the boilers. When temporary hardness is high, the soft scales are formed in the boilers causing corrosion.

7. Tentative water quality parameters for wet processing in textiles

pH → in the scale 6.5–7.5.
TDS → 250–300 ppm.
Color → Hazen No. 4.5–5.5.
Residue on ignition → 245–250 ppm.
Total Hardness → 25–30 ppm.
COD → to be considered as nil.
Turbidity → to be considered as nil.

Suspended Solids → to be considered as nil.
Copper (Cu) → to be considered as 0.01 ppm.
Iron (Fe) → to be considered as 0.01 ppm.
Chromium (Cr) → to be considered as 0.01 ppm.
Manganese (Mn) → to be considered as 0.05 ppm.
Aluminum (Al) → to be considered as 0.2 ppm.
Chloride (HCl) → to be considered as 125–150 ppm.
Sulphate (H2SO4) → to be considered as 125–150 ppm.
Nitrite → to be considered as nil.

8. Membrane based filtration in wet processing

The membrane based filtration is increasingly being used in the manufacturing processes of various process industries. Filtration by membranes based technology is considered as a profitable one than other types of traditional methods now a days. Their capacity to separate impurities and specific natural essences at low or ambient temperatures makes more usable. The selection of membrane is an important constituent for better functioning and optimum result of the process. There are many types of membrane that adapt to different use according to the level of filtration required. Ceramic membranes, spiral membranes, tubular membranes, stainless steel membranes, hollow fiber membranes and plate and frame membranes are the most common ones. Generally, filtration by membranes is a pressure technology that is utilized to separate various contaminated liquids. Its different types are ultrafiltration, microfiltration, nanofiltration and reverse osmosis.

Reverse osmosis is the processes of treatment of residual liquids particularly appropriate to dehydration, concentration/separation of substances. It is very helpful for the concentration of dissolved or suspended solids on the one hand and for obtaining a rejected liquid that comprise of a very low concentration of dissolved solids on the other hand.

Ultrafiltration is a process of exclusive division that is normally used to separate milk, whey and proteins. It paradigm suspended solids and solutes with a molecular weight of greater than 1000. The rejected liquid contains organic solutes having low molecular weight and salt.

Application of nanofiltration is normally work for demineralization, desalinization and also for color removal.

Microfiltration is a low pressure based flow through a membrane for the separation of colloids and suspended particles in the range of 0.05–10 microns. For the clarification of stock, fermentations and the clarification and recuperation of biomass, membrane based microfiltration is used.

9. Applications in industrial activities

Membrane based filtration is applied to various number of chemical processes industries. Filtration by membrane is very useful for food industries, specially in the dairy and sugar, pharmaceutical, biotechnological and chemical industries.

The utilization of various filtrations by membrane techniques in the food industries sets an infinite number of areas. The most popular include the concentration of egg whites, the elucidation and pre-concentration of various fruit juices, the concentration and natural process of the ashes of porcine, bovine or bone gelatin, the clarification of meat brine for the re-motion of bacteria and reprocess of the brine, the spatial arrangement of vegetable and plants such as canola, soy and oats and the separation of alcohol from wine and beer.

Dairy industry: In the manufacture of dairy ingredients, filtration by membrane is a valuable part of the production process. Its work can be divided into three major categories such as applications to milk, clarified cheese brine and whey.

Starch and sweetener industry: the main gain is the increase in the quality of the products, including the manufacturing and quality of corn syrups such as dextrose and fructose, the separation and property of rinse water from starch, the enhancement sub-product of dextrose, the depyrogenation of dextrose syrup and the division/application of maceration water.

Sugar industry: filtration by membrane can be used to purify unprocessed juice without using primary clarifiers, thus obviate many ambient problems and rising the quality and the execution of other traditional methods. The membranes can also purify, separate and concentrate various sugar solutions in the production activities and processes.

Chemical industry: many chemical processes use filtration by membrane. It works to desalinate, diafilter and purify dyes, pigments and optical brighteners. Filtration is also use to clean the waste water and rinse water currents. During the concentration and dehydration of minerals such as kaolin clay, titanium dioxide and calcium carbonate, during the clarification of caustic agents, the production of polymers or the recuperation of metals membrane based filtration process is used.

Pharmaceutical industry: the gathering of cells or the recovery of biomass is an essential part in the manufacturing process of fermentation, especially when manufacturing products such as antibiotics. Filtration process improves production as well as loss of the operator's workload and the maintenance reimbursement/cost. The membranes' filtration are also a standard part of the industrial manufacturing lines for enzymes, when concentrating enzymes prior to other processes.

Textile industry: many textile production processes exercise filtration by membrane to diafilter and purify dyes, desalinate, pigments and optical brighteners. It also uses to clean the waste water and rinse water currents, again for the concentration and dehydration of minerals such as titanium dioxide and calcium carbonate, the clarification of caustic agents, the manufacturing of polymers or the recovery of toxic metals such as Pb, Cr, Cu, etc.

10. Method

Samples were collected from different units in the study area at regular intervals during 2012–2013. Samples brought to the laboratory where they were analyzed using the standard method defined and suggested by American Public health Association (APHA, 1998). Temperature and pH were calculated by a mercury thermometer having scale from 0–100°C and with digital movable pH meter respectively at site. Total dissolved solids (TDS), dissolved oxygen (DO), salinity and turbidity were ascertained by Water Quality Analyzer PE- 371 (Systronic). Alkalinity of samples was calibrated by titrametric method. For analysis of COD, samples were stabilized by acidifying with H_2SO_4 below 2 and it was evaluated by dichromate titration method (APHA, 1998). The concentration of nitrate in the samples was ascertained by UV spectro photometric screening method with Zuconyl indicator. Sodium, potassium and calcium were calculated by Flame-photometric method. Heavy metal samples were analyzed after filtration by Whatman filter–paper no. 40 and then acidified samples were digested with concentrated HNO_3 (0.1%) acid. The metal ions were determined by atomic absorption spectrophotometer (OMA 300 process analyzer).

DOI: http://dx.doi.org/10.5772/intechopen.92812

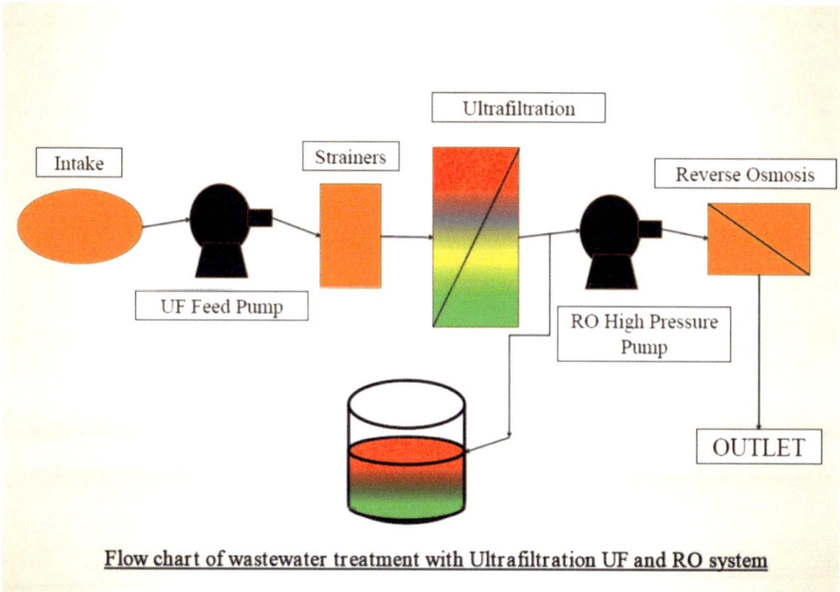

11. Results and discussion

The value of pH indicates the effluent before treatment is alkaline, substantial chemicals and dyes in solid form available in TDS, and trace metal concentration indicates that the raw effluent is hazardous for ecological systems in the region and required immediate treatment. The physicochemical parameters of wastewater generated from the bleaching and dyeing units after primary treatment found under limits as pH (7.7), BOD (20 mg/L), COD (120 mg/L), TDS (2234 mg/L), TSS (22 mg/L) by WHO (2003) and FAO (1985) [28–31] and trace

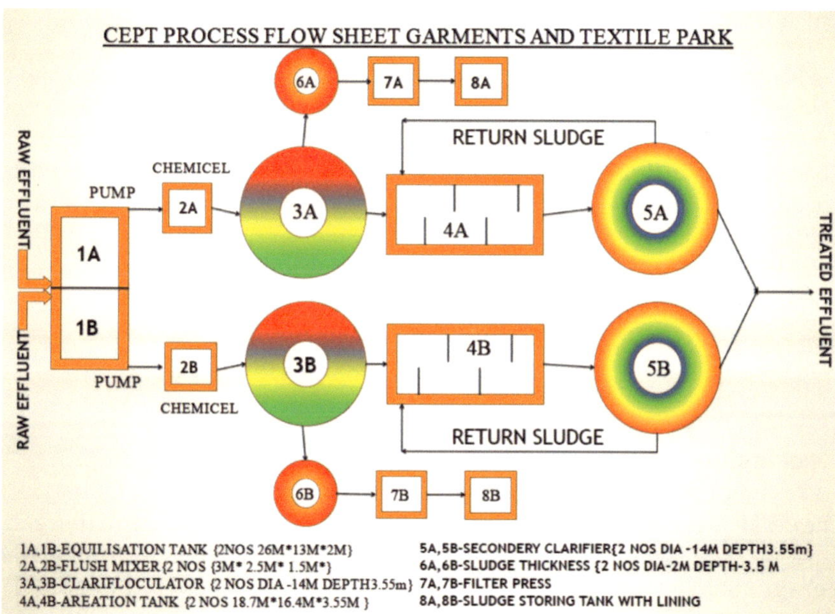

Figure 3.
CEPT PROSS flow chart: 500 MLD capacity.

Sr.no.	Parameters	Feed[a]	Permeate from	Concentrate from	Reverse osmosis
Ultrafiltration				Reverse osmosis	
1	Appearance	Clear	Clear	Clear	Muddy
2	pH	7.7.	7.7	6.0	7.8
3	Alkalinity	345(±15.2)	325(±14.1)	12(±1.4)	1100(±51.7)
4	Suspended solids	22(±1.6)	ND	ND	ND
5	Total dissolved solids	2234(±57.7)	2196 (±53.0)	40(±1.8)	7584(±195.4)
6	COD	120(±1.6)	20(±1.2)	ND	68(±2.3)
7	BOD	20(±1.2)	3(±1.2)	ND	11(±1.2)
8	Total Kjeldahl nitrogen	ND	ND	ND	ND
9	Phosphate	1.1(±0.08)	0.066(±0.05)	ND	0.22 (±0.06)
10	Sulphides	1.6(±0.20)	ND	ND	ND
11	Oil and grease	1.4(±0.20)	<1(±0.1)	ND	3.3(±0.40)
12	Chlorides	494(±29.1)	483(±28.4)	12(±0.80)	1653(±97.8)
13	Calcium	330(±11.1)	325(±11.0)	3(±1.1)	1125(±112.90)
14	Magnesium	164(±14.6)	143(±12.7)	2(±0.60)	493(±42.5)
15	Sulphate	350(±37.0)	307(±32.7)	ND	1070(±114.1)
16	Sodium	289(±14.0)	264(±12.9)	3(±0.9)	917(±122.9)
17	Potassium	15(±2.00)	<1(±0.20)	<1(±0.20)	1(±0.200)

All values are expressed in mg/L except pH; values in parenthesis are standard deviation; ND: not detectable.
[a]Tertiary treated effluent from feed tank of ATP, Source: Nandy et al. [32].

Table 1.
Efficacy assessment of advance treatment processes (ATP) after primary treatment.

metals Pb (0.33 mg/L), Cr (0.021 mg/L), Zn (0.54 mg/L), Ni (0.00 mg/L) and Cd (0.02 mg/L) will be cleaned after membrane based advance treatment processes (ATP) (**Figure 3** and **Table 1**).

The Common Effluent Treatment Plant (CETP) in Maheshtala cluster with advanced method of treatment containing of primary, secondary treatment, ultra-filtration and membrane based reverse osmosis, comparatively energy-efficient membrane based crystallization and distillation units and evaporation would yield premier quality water maintaining the qualifying standards of Central Pollution Control Board, New Delhi, India for industrial areas/ clusters [33].

12. Conclusion

The CETP plants can thus save surface and ground water from depletion and degradation. Transfer of contaminants to agri-horticulture produces through the food chain would be curtailed and thus human health risk would be minimized. The entire treatment and reusing of treated water will help to preserve 2000 MLD groundwater in the cluster area to save water and save life as the entire world is moving towards water paucity due to climate change.

Acknowledgements

The author thank to the students and teachers specially Sri Jayanta Kumar Biswas, Associate Professor, Department of Ecological Studies and International Centre for Ecological Engineering, University of Kalyani, Nadia, West Bengal, India for his guidance, cooperation and utilizing infrastructure and laboratories for the study.

The author did not receive any kind of funding from Government or any other non-government agencies.

Conflict of interest

The author does not have any types of financial and non-financial conflict of interest with any other person.

Abbreviation

APHA	American Public Health Association (1998). Standard methods for the examination of water and wastewater, WEF and AWWA, 20 th edition, USA
CPCB	Central Pollution Control Board (2007). Advance method of treatment of textile effluent
FAO	Food and Agriculture Organization, (1985). FAO guidelines of water quality for agriculture, Ministry of environment and forest, New Delhi
WHO	World Health Organization, (2003), Standard for quality of water for irrigation purpose
SWID	State water Investigation Department, Government of West Bengal

Author details

Biman Gati Gupta
Civil Engineering Department, Elitte College of Engineering (B. Tech), India

*Address all correspondence to: bimangupta@yahoo.in; civil.ece@petindia.com

References

[1] Olayinka KO, Alo BI. Studies on industrial pollution in Nigeria, the effect of textile effluents on the quality of ground water in some parts of Lagos. Nigerian Journal of Health and Biomedical Sciences. 2004;**3**(1):44-55

[2] Kannan V, Ramesh R, Sashikumar C. Study on ground water charectaristic and the effects of discharged effluents from textile units at Karur District, India. Journal of Environmental Biology. 2005;**26**(2):269-272

[3] Balachandra D, Sundarraj P, Ruthaevel MK, Kumarswamy K. An investigation of ground water quality and its suitability to irrigated agriculture in Coimbatore district, Tamilnadu, India, a GIS approach. International Journal of Environmental Sciences. 2010;**1**(2):176-190

[4] Hakanson L. An ecological risk index for aquatic pollution control a sedimentological approach. Water Research. 1980;**14**:975-1001. DOI: 10.1016/0043-1354(80)9014

[5] Bably P, Sangita K. Heavy metal pollution index of ground water of an abandoned open cast mine filled with fly ash: A case study. Mine Water and the Environment. 2008;**27**(4):265-267

[6] Prasad S, Mondal M, Aamodt K, Quintana AA, Achenbach R, Acounis S, et al. The ALICE experiment at the CERN LHC. Journal of Instrumentation. 2008;**3**(08):SO8002

[7] Aslan G, Akkaya C. Basic problems in groundwater sources and interactions between surface and groundwater. In: Ground Waters and Environment Symposium, 21-23 March 2001. Izmir. pp. 45-54

[8] Awomeso JA, Taiwo AM, Gbadebo AM, Adenowo JA. Studies on the pollution of waterbody by textile industry effluents in Lagos, Nigeria. Journal of Applied Sciences in Environmental Sanitation. 2010;**5**:353-359

[9] Baba A, Deniz O. The impact of open waste disposal site on surface and groundwater and investigation of groundwater quality of Canakkale City. Canakkale Onsekiz Mart University (Project number: 2001-b/04. p. 74); 2004

[10] Cuthbert JA. Hepatitis A: Old and new. Clinical Microbiology Reviews. 2001;**14**(1):38-58

[11] Chakraborty R, Vepuri V, Mhatre SD, Paddock BE, Miller S, Michelson SJ, et al. Characterization of a Drosophila Alzheimer's disease model: Pharmacological rescue of cognitive defects. PLoS One. 2011;**6**(6):e20799. DOI: 10.1371/journal.pone.0020799

[12] David PP, Dole JA. Explicit comprehension instruction: A review of research and a new conceptualization of instruction. The Elementary School Journal. 1987;**88**(2):151-165

[13] Uğurlu M. Effluents by the electrocoagulation method, Gazi University. Journal of Science. 2004;**17**(3):85-99

[14] Nair H, Nokes DJ, Gessner BD, Dherani M, et al. Global burden of acute lower respiratory infections due to respiratory syncytial virus in young children: A systematic review and meta-analysis. Lancet. 2010;**375**(9725):1545-1555. DOI: 10.1016/S0140-6736(10)60206-1

[15] Mohabansi NP, Tekade PV, Bawankar SV. Physico-chemical parameters of textile mill effluent, Hinganghat, Dist. Wardha (M.S.). Current World Environment: An International Research Journal of Environmental Sciences. 2011;**6**(1):165-168

[16] Ghugare SV, Chiessi E, Fink R, Gerelli Y, Scott A. Structural investigation on thermoresponsive PVA/poly (methacrylate-co-N-isopropylacrylamide) microgels across the volume phase transition. Macromolecules. 2011;**44**(11):4470-4478

[17] Verma AK, Dash RR, Bhunia P. A review on chemical coagulation/flocculation technologies for removal of colour from textile wastewaters. Journal of Environmental Management. 2012;**93**:154-168

[18] Joshi VJ, Santani DD. Physio-chemical characterization and heavy metal concentration in effluent of textile industry. Universal Journal of Environmental Research and Technology. 2012;**2**(2):93-96

[19] Rita K. Textile dyeing industry an environmental hazard. Natural Science. 2012;**4**(1):22-26. DOI: 10.4236/ns.2012.41004

[20] Patil A, Kim MS, Pinto SM, Getnet D, Nirujogi RS, Manda SS, et al. A draft map of the human proteome. Nature. 2014;**509**:7502, 575-7581

[21] Hiwat A, Teferi E. Urban Heat Island Effect of Addis Ababa City: Implications of Urban Green Spaces for Climate Change Adaption: Climate Change Adoption in Africa. Springer; 2014. pp. 539-552

[22] Garg R, Kumari R, Tiwari S, Goyal S. Genomic survey, gene expression analysis and structural modeling suggest diverse roles of DNA methyl transferases in legumes. PLoS One. 2014;**9**(2):e88947. DOI: 10.1371/journal.pone.0088947

[23] Arif S, Murat I, Almudi MDS, Nunes D, Bortolamiol-Becet NS, McGregor JMS, et al. Evolution of mir-92a underlies natural morphological variation in Drosophila melanogaster. Current Biology. 2013;**23**(6):523-528

[24] Wazir A. Application of Bacterial Pigments as Colorant: The Malaysian Perspective. Springer; 2012. pp. 1-75

[25] Hu Y, Tang HB, Liu NN, Tong XJ, Dang W, Duan YM, et al. Telomerase-null survivor screening identifies novel telomere recombination regulators. PLoS Genetics. 2013;**9**(1):e1003208

[26] Goyal MK, Chauhan A, Chauhan P. GC-MS technique and its analytical applications in science and technology. Journal of Analytical and Bioanalytical Techniques. 2014;**5**:222. DOI: 10.4172/2155-9872.1000222

[27] Youssef MA, Eissa MA. Heavy metals accumulation in the edible vegetables grown on contaminated soils. Egyptian Journal of Soil Science. 2015;**5**:1-14

[28] Ranganathan K, Karunagaran K, Sharma DC. Recycling of wastewaters of textile dyeing industries using advanced treatment technology and cost analysisd. Case studies. Resources, Conservation and Recycling. 2007;**50**:306-318

[29] Singhal V, Gupta KC. Visit Report of Effluent Treatment Plants/ CETP in the Country. Jaipur: Rajasthan State Pollution Control Board; 2010

[30] Rao AV. Textile effluent, biological treatment. Journal of Environmental Research And Development. 2005;**12**:133-140

[31] Ghaly AC, Benden C, Hyden K. Effective treatment of textile effluent by conventional methods. Environmental & Resource. 2014;**21**:211-213

[32] Nandy T, Manekar P, Dhoparkar R, Pophali G, Devotta S. Water conservation through implementation of ultrafiltration and reverse osmosis system with recourse to recycling of effluent in textile industry—A case study. Resources,

Conservation and Recycling. 2006. DOI: 10.1016/J.resconrec.2006.08.0046

[33] Gupta BG, Agrawal KM, Biswas JK. Conservation of ground water at Maheshtala bleaching and dyeing cluster, a populated area in West Bengal, India by implementing ultra filtration and reverse osmosis based effluent treatment plant—A case study. Journal of The Institution of Engineers Springer. 2017. DOI: 10.1007/s40030-018-0314-7

Chapter 6

Solid Oxide Steam Electrolyzer: Gas Diffusion Steers the Design of Electrodes

Jonathan Deseure and Jérôme Aicart

Abstract

The hydrogen production by SOECs coupled with renewable energy sources is a promising route for the sustainability *hydrogen economy*. Multiphysics computing simulations appear to be the most efficient approaches to analyze the coupled mechanisms of SOEC operation. Using a relevant model, it is possible to predict the electrical behavior of solid oxide electrodes considering the current collector design. The influences of diffusion and grain diameter on cell performances can be investigated through 2D simulations, current–voltage characteristics, and current source distribution through electrodes. The simulation results emphasize that diffusion is linked to a relocation of the reaction away from the interface electrolyte/electrode, in the volume of the cathode. Furthermore, the current collector proves itself to be a great obstacle to gas access, inducing underneath it a shortage of steam. Inducing gradients of grain diameters in both anode and cathode drives the current sources to occur close to the electrode/electrolyte interface, thus decreasing ohmic losses and facilitating gas access. This approach shows the crucial importance of cathode microstructure as this electrode controls the cell response.

Keywords: hydrogen production, steam electrolysis (SOEC), electrochemical modeling

1. Introduction

Usual electrolyzers employ aqueous electrolytes (alkaline water electrolyzer), and the major drawback of the electrolytic hydrogen production is its high cost in comparison with the steam-methane reforming process [1]. In addition to alkaline-based electrolysis, there are mainly two technologies of electrolyzer, which are currently considered, one based on proton-exchange membrane (PEM) and the other based on solid oxide (SOECs). The electrochemical decomposition of steam into hydrogen and oxygen offers two advantages over the low-temperature process currently in use. High temperature electrolysis (HTE) is more efficient than traditional room-temperature electrolysis for many reasons. First, some of the energy is supplied as heat, which is cheaper than electricity. Secondly, at high temperature the electrolysis is less energy consuming due to the low theoretical decomposition voltage. In the 1980s, solid oxide fuel cell (SOFC) developments allowed steam electrolysis investigation [2]. Thus, several experimental studies on high-temperature electrolysis (HTEs) using practical SOFC cells [3] have been carried out. These experiments showed encouraging results. The solid oxide electrolysis cell (SOEC) is the reverse operation of the same SOFC cell structure. Both are composed of two porous composite ceramic electrodes surrounding a gas-tight electrolyte. SOECs can rely on the interest that SOFCs have received for the past decades and thus utilize similar technology and materials. Water is reduced at the cathode Eq. (1), releasing hydrogen and oxygen ions. After crossing the electrolyte, the ions are then being oxidized to form oxygen within the anode Eq. (2). Both half reactions are balanced by general Eq. (3):

$$H_2O + 2e^- \rightarrow H_2 + O^{2-} \tag{1}$$

$$O^{2-} \rightarrow \tfrac{1}{2}O_2 + 2e^- \tag{2}$$

$$H_2O \rightarrow H_2 + \tfrac{1}{2}O_2 \tag{3}$$

Due to the high operating temperature, SOECs do not need expensive catalysts, but these must meet strict thermal, chemical, and structural requirements imposed by temperature, and hydrogen and oxygen partial pressures. However, performance still remains limited in electrolysis mode compared to fuel cell mode [4].

Comparing to PEM electrolyzers for which carbon monoxide is a poison, SOECs offer the advantage to allow the co-electrolysis of water and carbon dioxide to produce syngas [5, 6]. According to AlZahrani and Dincer [7] a SOEC system can achieve energy and exergy efficiencies of 53 and 60%, respectively. However, the high operating temperature (>1000 K) is still considered the major limiting factor of these device. Jiang [8] has shown that the hydrogen production by SOECs coupled with renewable energy sources is a promising route for the sustainability of energy in the future. The solution of a real commercially competitive SOC technologies is the materials: reliability and stability of the electrode and electrolyte materials under the reversible electrolysis and fuel cell operation modes. The reversible fuel cells have the ability to switch between electrolysis cell and fuel cell modes, and it is one of the foremost features that facilitate storing/generating energy in a cost-effective manner. The optimized parameters on the designs of fell cells or of electrolyzer, solely, would not necessarily result in high performance of regenerative devices because these devices differ in electrode kinetics, gas environment, heat generation, and chemical stability. It is well known that high-temperature operation of SOECs offers inherent advantages, in terms of thermodynamics and kinetics compared with low-temperature electrolysis. In this context reversible solid oxide cells (RSOCs) are still at an early stage of development [9, 10]. Unfortunately, there is a general consensus that the performance and stability of SOECs are inferior to those of SOFCs [11], which is mainly due to the high-temperature operation.

Computing simulation appears to be one of the most efficient approaches to analyze the coupled mechanisms of SOEC operation. It can predict the SOEC behavior under various operating conditions. Mathematical modeling is an essential tool in the design of SOEC cells, as it is important to understand the limiting process of steam electrolysis. Recent literature shows a significant research and development effort focusing on the modeling of SOEC. The models developed by Udagawa et al. [12, 13] strive to describe all significant processes affecting the performance of a unit cell. These authors proposed one-dimensional or pseudo-2D simulations based on a planar geometry and taking into account mass transport. Ni et al. [14, 15] have described the mass transport within the electrode along with the electrochemical kinetics. The principal results of these investigations [16, 17] lead to a parametric control of the SOEC operation. In addition, Jin and Xue [18, 19] have developed a 2D model for a planar SOEC. The simulation results lead to a better understanding of the internal mechanisms for regenerative SOFCs. This model has been used to study the delamination phenomenon at the oxygen electrode/electrolyte interface. Few micro-modeling investigations are applied to SOEC, and relevant electrochemical models are required to improve the micro-scale predictions [20–23]. Nevertheless, the results of SOFC micro-modeling can be employed to appreciate, for example, particle size, graded or homogeneously distributed porosity, or composition influence on electrode performance.

In the subsection below a multiphysics model of SOEC has been built using the commercial software Comsol Multiphysics®. Electrochemical reactions within porous electrodes are described using the Butler-Volmer equation. Modeling is based on solving conservation equations of mass, momentum, and charge balance. Simulations allow the calculation of gas concentration, current density, and potential distribution within the electrodes (i.e., interconnects and electrodes). These simulations establish how porous electrode performance is affected by current collectors and electrode microstructural parameters. The model is then applied to engineer a design of the electrode structure or current collector configurations.

2. Model equations

In the present model, mass and charge transport phenomena coupled with chemical and electrochemical reactions have been investigated within the SOEC cell. This mathematical approach is based on classical SOFC assumptions, and thus the model should depend on operating conditions, intrinsic conductivities of materials, and geometric parameters such as porosity or grain size [21–22]. Additionally, according to the 2D model of Kenney and Karan [23], the interconnects play a critical role. Thus, a 2D approach was performed in this work. In this study, a finite element method has been used to solve mass and charge balances including transport through porous media and electrochemical reactions within the porous electrodes. The set of resulting conservation equations has been solved using the commercial software Comsol Multiphysics®. In this computational approach, steady-state conditions have been imposed. This model of SOEC is based on the following assumptions:

- Perfect current collectors (equipotential surface with perfect contact)
- No contact resistances
- Constant pressure
- Negligible convective flow through the electrodes
- Ionic and electronic conductivities depending solely on temperature

Equations are detailed in this section one balance at a time.

2.1 Charge balance

The electrode material is a mixed electronic and ionic conductor. For modeling purposes, this electrode is considered as a porous gas diffusion electrode wherein the electrochemical reaction occurs at the triple phase boundary, i.e., at the interface between the electronic conductor, ionic conductor, and gas phase. The current in a porous electrode can be split into two parts: one part flowing through the ionic phase and the other through the electronic phase of the porous matrix. During electrochemical reactions, electrons are then transferred from the ionic phase to the electronic phase or vice versa. The transport of each kind of charges (e^-, O^{2-}) can be described using Ohm's law. To account for charge transfer between electronic and ionic materials, a current source term $Q_{i,a/c}$ (A m^{-3}) is employed in the charge balance in Eq. (4):

$$-\nabla.\left(\sigma_{S/M,a/c}^{eff}\nabla\phi_{S/M}\right) = Q_{i,a/c} \tag{4}$$

The effective conductivity ($\sigma_{S/M,a/c}^{eff}$) depends on the material and the microstructure of each electrode. Their values can be computed using Eq. (5) [20], where S and M subscript are, respectively, ionic material and electronic material.

$$\sigma_{S/M,a/c}^{eff} = Y_{S/M,a/c}\frac{\left(1-\varepsilon_{a/c}\right)}{1.6}\sigma_{S/M,a/c} \tag{5}$$

The bulk conductivities and molar fractions that have been used throughout this work are gathered in **Table 1**. Nickel has been used as the electronic material at the cathode, whereas the anode was made of LSM-type perovskite [24, 25]. YSZ allows the transport of ions in both electrodes and the electrolyte. Within the electrolyte ceramic membrane, there are no current sources. Therefore, pure and dense YSZ electrolyte is considered (σ_{YSZ}) in Eq. (4). Similarly, current collectors are assumed to be ideal electronic conductors. The charge balance ensures that the current produced at the cathode is consumed at the anode. Additionally, in each electrode, the electronic current is the opposite of the ionic one. According to Costamagna et al. [26], the current source terms $Q_{i,a/c}$ can be described by the classical Butler-Volmer expression Eqs. (7) and (8), and the current sources are expressed as follows:

$$Q_{i,a/c} = \pm i_{bv,a/c} \tag{6}$$

The selected parameters of Butler-Volmer equation remain valid at high fuel utilization and low value of hydrogen concentration corresponding to the SOEC mode. The current sources and therefore the expression of the electrochemical reactions are expressed by the following Butler-Volmer Eqs. (7) and (8), F being the Faraday's constant (F = 95,485 C mol^{-1}) and R the ideal gas constant (R = 8.314 J mol^{-1} K^{-1}):

$$i_{bv,c} - i_{0\_c}\left(\frac{C_{H_2}}{C_{H_2}^0}\exp\left(\frac{2\alpha_c F\eta_c}{RT}\right) - \frac{C_{H_2O}}{C_{H_2O}^0}\exp\left(\frac{-2(1-\alpha_c)F\eta_c}{RT}\right)\right) \tag{7}$$

Parameter	Value
Nickel electronic conductivity, σ_{Ni} [S m^{-1}] [24]	4.5×10^5
LSM electronic conductivity, σ_{LSM} [S m^{-1}] [24]	1.6×10^5
YSZ ionic conductivity, σ_{YSZ} [S m^{-1}] [27]	$\sigma_{YSZ} = 0.334 \times 10^5 \exp\left(-10300/T\right)$
Cathodic volume fraction of nickel, $Y_{Ni,c}$ [24]	0.4
Anodic volume fraction of LSM, $Y_{LSM,a}$ [24]	0.5
Cathodic volume fraction of YSZ, $Y_{YSZ,c}$ [24] Anodic volume fraction of YSZ, $Y_{YSZ,a}$ [24]	0.6 0.5

Table 1.
Conductivities and molar composition of the SOEC.

$$i_{bv,a} = i_{0\_a}\left(\exp\left(\frac{2\alpha_a F\eta_a}{RT}\right) - \frac{C_{O_2}}{C^0_{O_2}}\exp\left(\frac{-2(1-\alpha_a)F\eta_a}{RT}\right)\right) \tag{8}$$

where $\eta_{a/c}$ are the overpotentials defined as the difference between the electronic potential on one hand and the ionic and equilibrium potentials on the other hand, displayed in Eq. (9). In the present model, hydrogen electrode is assumed to have a thermodynamic potential equal to 0 V:

$$\eta_{a/c} = \phi_{electronic,a/c} - \phi_{ionic,a/c} - E'^{0}_{a/c} \tag{9}$$

According to [24], the influence of the electrode microstructure on the exchange current densities can be expressed by Eq. (10):

$$i_{0,a/c} = i^{ref}_{0,a/c}\left(1 - \frac{\varepsilon_{a/c} - 0.26}{1 - 0.26}\right)\left(\frac{dg^{ref}}{dg_{a/c}}\right)^3 \tag{10}$$

Laurencin et al. [16] suggested that Butler-Volmer expression is suitable to describe electrochemical reaction involving the exchange of 2 electrons. The values for electrochemical geometric symmetric coefficients α_c and α_a are usually considered to be close to 0.5 in the literature [26, 28]. The exchange current densities $i^{ref}_{0,a/c}$ are critical parameters to describe the current generated by the cell. Both parameters take into account the virtual specific surface area linked to the vicinity of the triple point boundaries (TPB) where the charge transfer occurs. These parameters have been deduced from previous work [16, 24] in order to observe usual SOEC performance, and are gathered in **Table 2**.

Finally, the electrolyte material is YSZ, a suitable ionic conductor at high temperatures. The electrolyte potential is thus expressed by a classical Ohm's law Eq. (4) without any current source term ($Q_{S,a/c} = 0$) within the electrolyte.

The Nernst law Eq. (12) is used to assess the open circuit voltage (OCV) at operating temperature T of 1173.15 with an E^0 of 1.1 V which, in turn, is used to obtain the cell voltage through Eq. (11) with V_{pol} the parametric input of the polarization computation. Vpol has been defined in order to stabilize the numerical convergence near the OCV operating point. The partial pressures are according to the inlet feed.

$$V_{cell} = E_{eq} + \eta_{cell} = E_{eq} + \left(V_{pol} - \phi_{electronic,c}\right) \tag{11}$$

$$E_{eq} = E^0 + \frac{RT}{2F}\ln\left(\frac{P_{H_2,c}P^{1/2}_{O_2,a}}{P_{H_2O,c}}\right) \tag{12}$$

Parameter	Value	References
Cathodic exchange current density, $i^{ref}_{0,c}$ [A m^{-3}]	4×10^8	[16, 24]
Anodic exchange current density, $i^{ref}_{0,a}$ [A m^{-3}]	4×10^7	[16, 24]
Symmetrical factors, $\alpha_{a/c}$ [−]	0.5	[24]
Reference grain diameter, dg^{ref} [μm]	3	This work

Table 2.
Parameters used to apply the Butler-Volmer equation.

Parameter	Value
Electrodes porosity, $\varepsilon_{a/c}$ [−]	0.37
Electrodes grain diameter, $dg_{a/c}$ [μm]	3
Electrodes tortuosity, $\tau_{a/c}$ [−]	4.8

Table 3.
Microstructural parameters used in this study.

2.2 Mass balance

As stated in the assumptions of the model, since electrodes are porous media with low permeability, mass transport is only due to diffusion process. Binary gas interaction and pore wall effect should be taken into account using the following general expression of the mass balance:

$$\nabla.\left(-D_k^{eff}\nabla C_k\right) = \Gamma_k \quad (13)$$

where Γ_k is the mass source term, directly linked to the current sources $Q_{a/c}$ by Faraday's law as expressed below:

$$\Gamma_{O_2} = \frac{i_{bv,a}}{4F} \quad (14)$$

$$\Gamma_{H_2} = -\Gamma_{H_2O} = \frac{i_{bv,c}}{2F} \quad (15)$$

The binary diffusion coefficient of water in hydrogen was computed through Eq. (26), presented in appendix. Thus, at the cathode, only hydrogen and steam are taken into account. The effective diffusion coefficient takes into consideration the microstructure of the electrode according to Eq. (16) [24]:

$$D_{H_2O\_H_2}^{eff} = D_{H_2O\_H_2}.\varepsilon_C^{\tau_C} \quad (16)$$

In the porous media of each electrode, and because of the presence of small pore size, the model integrates the Knudsen diffusion process, which coefficient is expressed by Eq. (17) [20]:

$$D_{H_2O}^{K} = \frac{dp_C\varepsilon_C}{3\tau_C}\sqrt{\frac{8RT}{\pi M_{H_2O}\times 10^3}} \quad (17)$$

Table 3 gathers grain diameters ($dg_{a/c}$) and tortuosity ($\tau_{a/c}$) values. Pore sizes are a function of both parameters, as displayed by Eq. (18) [20]:

$$dp_{a/c} = \frac{2\varepsilon_{a/c}dg_{a/c}}{3\left(1-\varepsilon_{a/c}\right)} \quad (18)$$

Usually, the dusty gas model considers Maxwell and Knudsen diffusion to describe the gas flow through a porous media. It was possible to suggest an approach equivalent to Fick's law (Eq. (13)) by considering an effective equivalent diffusion coefficient [24], illustrated by Eqs. (19) and (20), with β_c being the Fick diffusion form coefficient:

$$D_{H_2O}^{eff\_F} = \frac{1}{\frac{1}{D_{H_2O}^{K}} + \left(1 - \frac{\beta_c C_{H_2O}}{C^{tot}}\right)\frac{1}{D_{H_2O\_H_2}^{eff}}} \tag{19}$$

$$\beta_c = 1 - \sqrt{M_{H_2O}/M_{H_2}} \tag{20}$$

In the set of mass balance equations, the ideal gas law is applicable Eq. (21). This work has been done at a working temperature T of 1173.15 K and a pressure P of 10^5 Pa:

$$C^{tot} = \frac{P}{RT} \tag{21}$$

Similar modeling has been performed at the anode side to describe mass flow through the porous anode. O_2 and N_2 are substituted to H_2O and H_2, respectively. The binary diffusion coefficient $D_{O_2\_N_2}$ is presented in Annex under Eq. (27). Eqs. (22)–(24) display the Knudsen diffusion coefficient, the effective diffusion coefficient, and the Fick diffusion form coefficient, respectively:

$$D_{O_2}^{K} = \frac{dp_A \varepsilon_A}{3\tau_A}\sqrt{\frac{8RT}{\pi M_{O_2} \times 10^3}} \tag{22}$$

$$D_{O_2}^{eff\_F} = \frac{1}{\frac{1}{D_{O_2}^{K}} + \left(1 - \frac{\beta_a C_{O_2}}{C^{tot}}\right)\frac{1}{D_{O_2\_N_2}^{eff}}} \tag{23}$$

$$\beta_a = 1 - \sqrt{M_{O_2}/M_{N_2}} \tag{24}$$

3. Simulation conditions

The commercial software Comsol Multiphysics® has been used to investigate the behavior of a simplified serial repeating unit (SRU) of SOEC. The modeled geometry will thus include both electrodes, the electrolyte, and the current collectors. The resulting set of conservation equations is solved using the commercial software. This section will focus on the description of how the model was implemented.

3.1 Geometry

This study objective has been to investigate the influence of the electrode microstructure considering a realistic geometry of the SRU. **Figure 1** displays the geometry that forms the basis of the model. For simplicity's sake, a two-dimensional model of the SRU was used according to its cross section. The cell dimensions are gathered in **Table 4**. A mapped mesh has been used to obtain a reasonable number of degrees of freedom allowing calculation convergence within an acceptable computation time. Its parameters are also listed in **Table 4**. The studied geometry is a cross section of SRU which is perpendicular to the main direction of the gas flow. Considering that channels allow an ideal distribution of reactants on each electrode, it is possible to consider them as boundary conditions. This means that the produced hydrogen and oxygen are perfectly collected and exhausted from the SRU by the gas manifolds. Therefore, constant gas concentrations within the gas channels have been considered, in order to use this value as boundary conditions of mass balance.

3.2 Boundary conditions

This section presents the boundary conditions used to solve each balance. The symmetry existing in the geometry allows the solver to consider half of the mesh. The following **Figure 2** summarizes the different boundary conditions defined. The present multiphysics problem uses partial derivative equations that need boundary

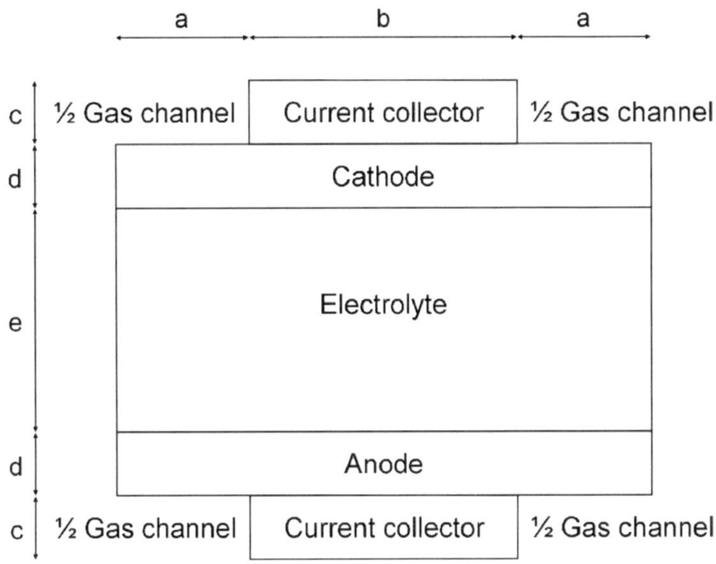

Figure 1.
Model geometry.

SRU dimensions		Mesh parameter	
a [mm]	1	Degrees of freedom	63,982
b [mm]	2	Number of mesh point	13,757
c [mm]	1	Number of boundary elements	1352
d [μm]	40	Minimum element quality	0.5505
e [μm]	100		

Table 4.
Cell dimensions and computational parameters.

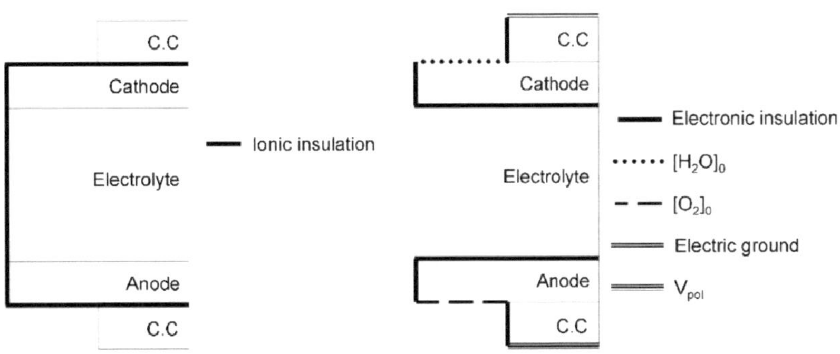

Figure 2.
Type and localization of the model's boundary conditions.

Boundary condition	**Expression**
V_{cell}	$\phi_{elec} = V_{pol}$
Ground	$\phi_{elec} = 0$
Electric insulation	$n.i = 0$
Continuity	$n.(i_1 - i_2) = 0$
Gas insulation	$n.(-D\ \nabla C_i + C_i\ u) = 0$
Cathode inlet composition	$[H_2O]_0$ = 10.09 mol m^{-3} (90%$_{molar}$) $[H_2]_0$ = 1.12 mol m^{-3} (10%$_{molar}$)
Anode inlet composition	$[O_2]_0$ = 2.35 mol m^{-3}

Table 5.
Mathematical expressions and numerical values of the boundary conditions.

conditions to be solved, such as electric and ionic charge balances Eq. (4) or the mass balance Eq. (13). Such conditions are expressed by the set of equations presented in **Table 5** that are to be satisfied.

4. Results and discussion

The simulations obtained with the multiphysics model previously described have been developed to investigate the impact of diffusion on the SOEC performance and to quantify the location of current sources within functionally graded electrodes.

4.1 Control by diffusion

The diffusion phenomenon was investigated via four simulation cases, referred to with letters A to D. A is the reference case, based on the set of parameters and geometry as described in **Tables 2–4**. This set was modified to give simulations B, C, and D. Case B considers a perfectly collected current without the collector pin: the boundary conditions of mass balance and electronic charge balance are merged. The effect of electrode thickness on gas diffusion and reaction distribution within the electrode was studied with case C considering an electrode thickness of 80 μm instead of the previous 40 μm (case A). Finally, case D displays a steam diffusion coefficient 10 times lower than in the standard case A. **Figure 3** shows the results obtained for case A (left), taken as a reference for the discussion, and case B (right), where the collector pin has been removed.

The distribution of current sources (i_{bv}) appears to be neither continuous nor homogeneous. **Figure 4** exhibits the steam concentration distribution along the electrode at varying abscissa. Whereas high current densities are reached in the vicinity of the interfaces electrode/electrolyte under the gas canals, a low fraction of the total current is produced under the current collector. The water concentration under the cathodic current collector is very low, and this shortage is close to depletion.

Consequently, most of the current source terms are located below the gas channel, and these sources are rather small below the current collector. Since most of the current is generated below the gas channels, the convergence of the electrons toward the collector pin causes hot spots where high current densities are observed. To separate the role of diffusion from other phenomena, such as conduction of charged species, additional simulations were performed. In order to complete the investigation of diffusion process, additional simulation cases B to D were

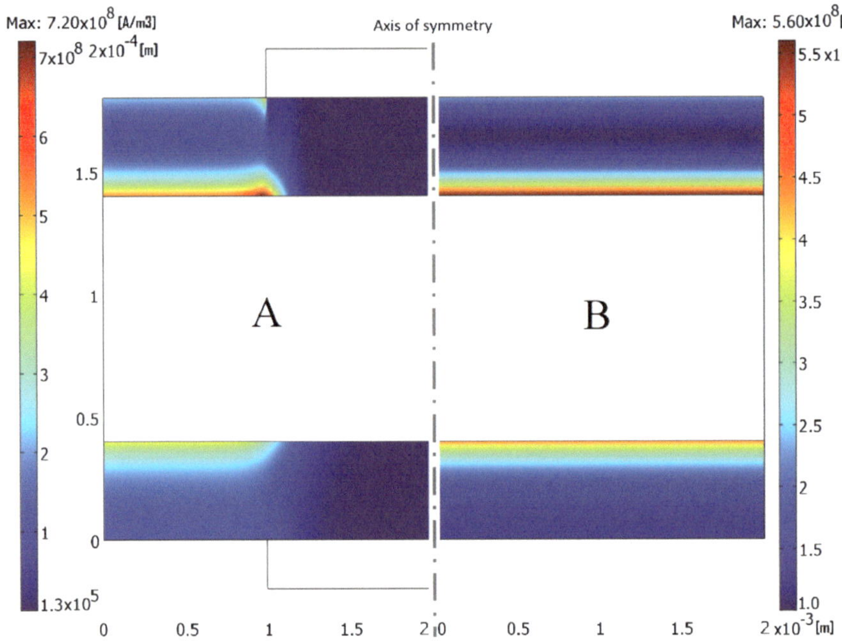

Figure 3.
Current sources (A m^{-3}) through SOEC cell for a cell potential close to 1.3 V, for case A (left, a) and case B (right, b).

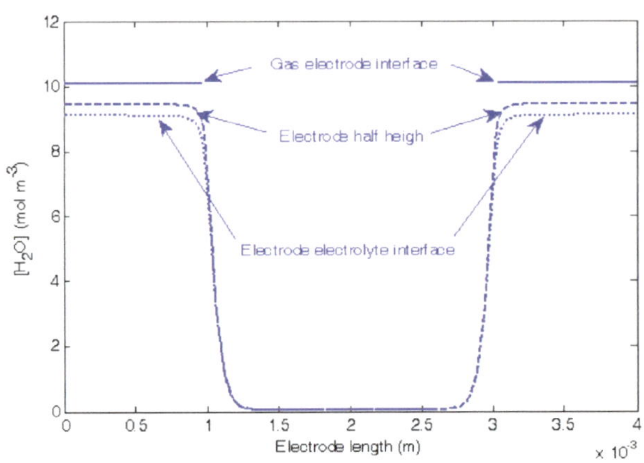

Figure 4.
Steam concentration (mol m^{-3}) through the SOEC cathode for a cathodic overpotential equal to 0.5 V (cell voltage = 1.3 V) for case A.

performed. Consequently, as seen in **Figure 3b**, no water depletion is observed in case B when considering a perfectly collected current without the collector pin. The current–voltage characteristics for cases A, B, C, and D are presented in **Figure 5**.

As expected, a homogeneous current collecting significantly improves the performance. Gaseous reactants and products are not impeded by the collector pin anymore. Decreasing the water diffusion coefficient induces a large shift of the current-tension characteristic toward lower currents and thus increases concentration overpotentials. According to Juhl et al. [29], increasing the electrode thickness improves the electrode performance for LSM/YSZ composite cathode of SOFC using thicknesses between 2 and 12 μm. Nevertheless, Virkar et al. [22] have computed that SOFC cathode overpotential increases or decreases with electrode thickness depending on the ionic conductivity of the composite. In the present simulations, both conductivities (electronic and ionic) and gas transport have been computed, and the competition between gas, ions, and electrons was investigated. The simulation results highlight a complex distribution of electrochemical active areas (see **Figure 6**). Due to the competition effect between gas (water) diffusion and ionic charge transport occurring in "countercurrent," the current source terms are located close to both interfaces. In order to obtain better comprehensions, the current sources distributions have been analyzed. **Figure 6** presents diagrams of Faradic currents (A m^{-3}) along the cathode thickness for the studied cases. The electrode thickness has been divided into 10 layers from the electrolyte/cathode interface to the cathode/gas channel interface, and the ratio of the current sources to the whole current for each layer is plotted versus the electrode thickness. The results are expressed in percentage of the total current generation. Since the most interesting phenomena occurs under the edges of the collector pin, the current source terms are considered at abscissa x = 1 mm.

The competition between the transport of ions and the diffusion of gases is highlighted. Electron transport is not a limiting step as the electronic conductivity is about 10^4 times higher than the ionic one. When the water diffusion coefficient decreases, the competition between O^{2-} and gas transport is distorted, and a relocation of current sources toward the cathode/gas channel interface occurs (**Figure 6**). Therefore, gas access implies a higher reactivity of the electrode/gas channel interfaces than expected.

On all 40 μm thick electrodes, all layers exhibit a reactivity higher than 5% of the total generated current. However, the 80 μm electrode generates roughly 65% of the total current within the first 24 μm of electrode thickness and 30% within the last 16 μm (**Figure 6c**). In other terms, 50% of the electrode thickness is responsible for 95% of the current generation. The thinner the electrode, the more homogeneous the current sources seem to be.

Figure 7a shows the current source generation along the 80 μm electrode thickness located in the middle of the gas channel (abscissa x = 0 mm). It can be seen that more than 70% of the total current generation comes from the first 30% of the electrode thickness. That observation is coherent with Hussain et al. [30] who consider that the electrochemical reaction is occurring exclusively at the interface electrode/electrolyte. **Figure 7a, b** show that increasing electrode thickness leads to

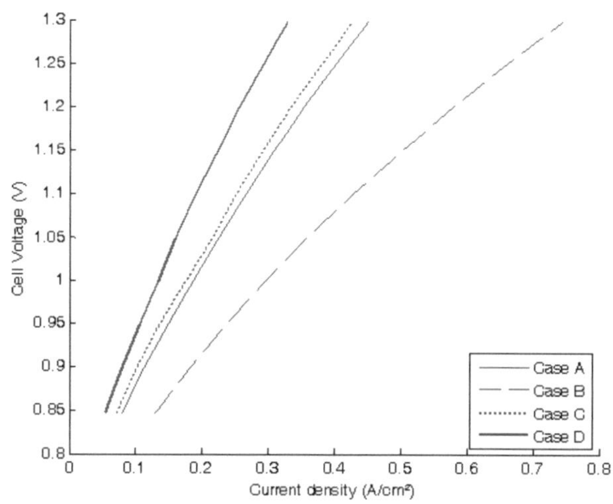

Figure 5.
Polarization curves for cases A, B, C, and D.

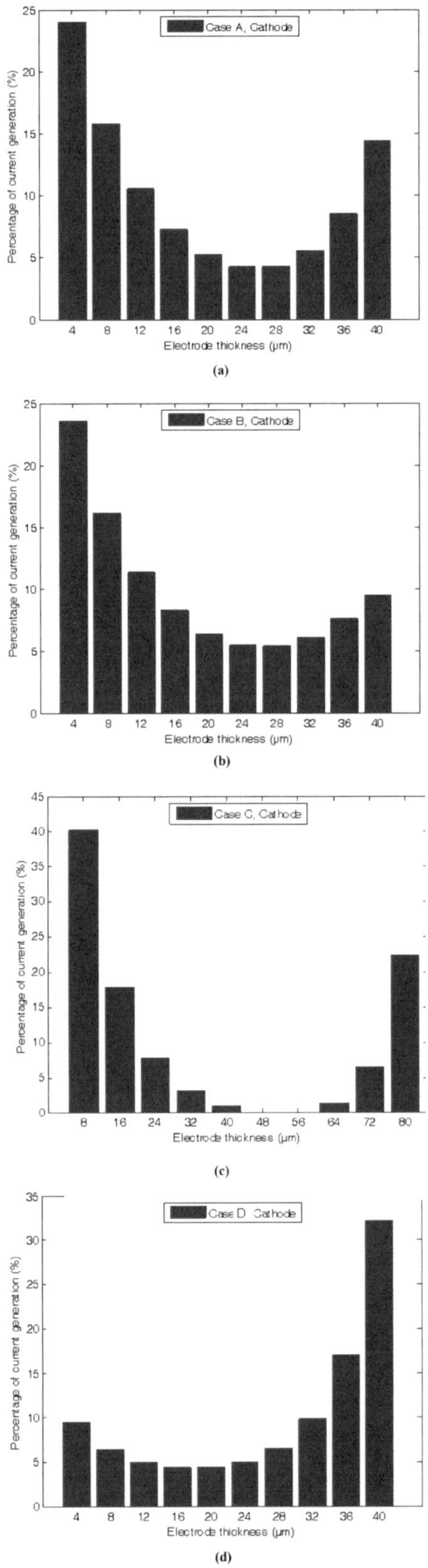

Figure 6.
Current sources distribution through cathodes for cases A (a), B (b), C (c), and D (d), at abscissa separating the gas channel and the current collector (x = 1 mm and cell voltage = 1.3 V).

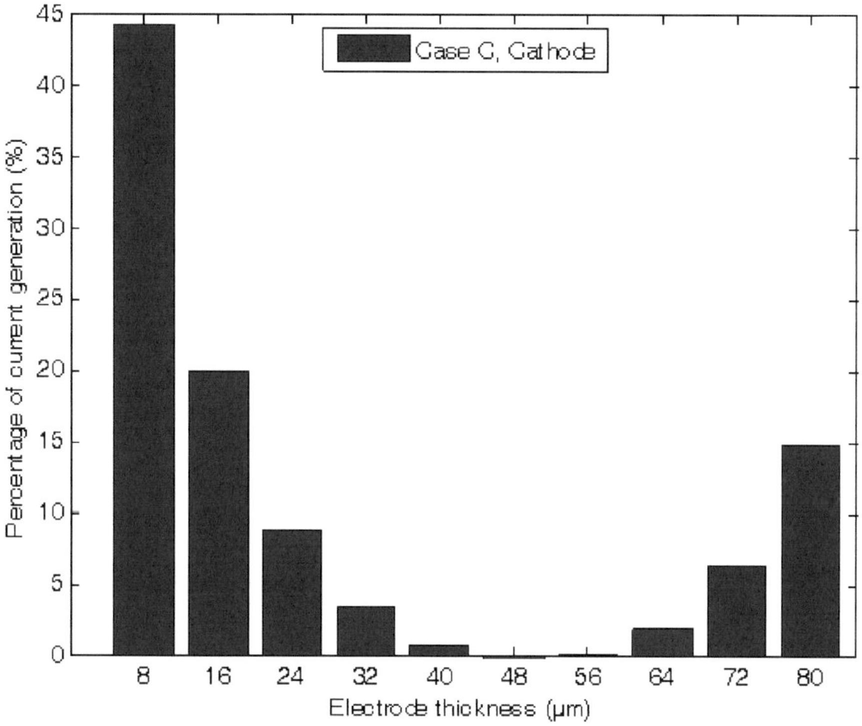

(a)

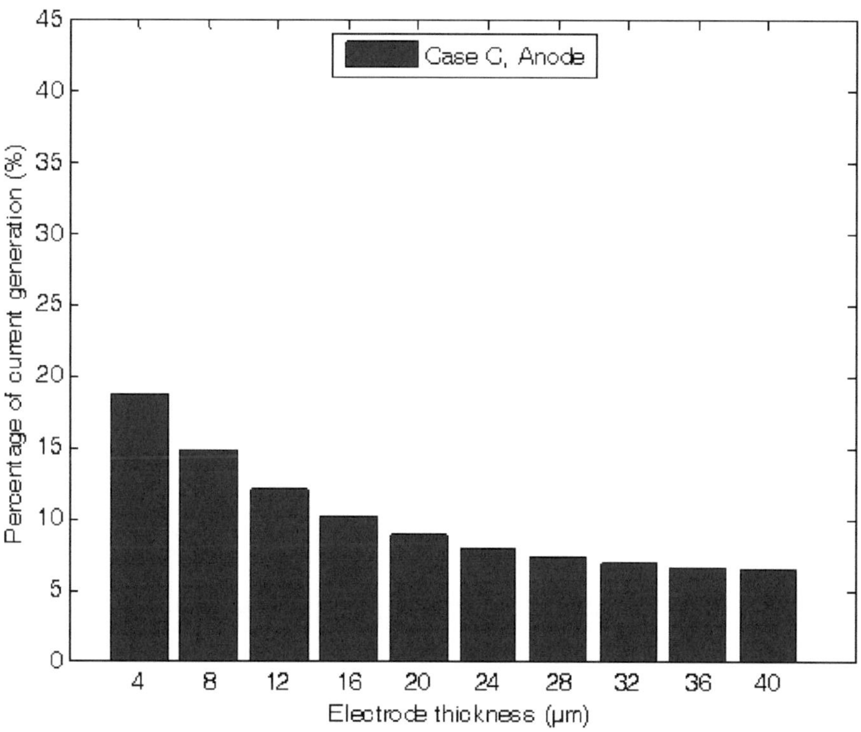

(b)

Figure 7.
Current sources distribution though case C (80 μm electrode) cathode (a) and anode (b) at abscissa located in the middle of the gas channel (x = 0 mm and cell voltage = 1.3 V).

current being generated closer to the collector pin. This is compensated by a higher reactivity in the interface vicinity. Those edge effects, obvious in all cases, can be attributed to the electronic ohmic drop. Moreover, the ionic ohmic drop also controls the distribution of anodic current sources. **Figure 7b** displays the anodic equivalent to **Figure 7a**. Contrary to the cathode, no control of the faradaic currents by gas access exists at the anode, since the gas is being produced. Nevertheless, both electrodes display similar current sources distributions. Such observation remains valid for all simulations. Consequently the anodic current sources distribution is driven by electrical cathodic behavior.

4.2 Influence of graded grain diameter

It is generally accepted that the performances of composite electrodes as well as graded cathodes in SOFCs are largely governed by TPB number, mass transport, and ohmic drop. It is well known that the polarization resistance decreases [31, 32] when using graded electrodes. The improvement of the microstructure is one of the key parameters to reach high electrochemical performances.

In the context of current collecting considerations, this work has emphasized the effects of functionally graded electrodes on overall cell behavior (i.e., current collectors/electrodes/electrolyte). Thus, the influence of grain size distribution was investigated via the simulations gathered in **Table 6**. A gradient of grain diameter distribution was introduced, leading to either the cathode or the anode presenting several layers of different grain diameters. The effects of such a change in microstructure on the currents obtained and their distribution were investigated. In this work, electrodes are constituted of two layers, each presenting a specific grain size. Their responses were modeled and the results analyzed. For each cathode and anode, five simulations were performed, referenced from I to V, with the subscripts "a" and "c" referring to anode and cathode, respectively. One goal of this work was to investigate the influence of the grain diameter on the reaction location and the cell performances. However, as Eq. (10) shows, the exchange current density is modified by the microstructure. To be able to compare the currents obtained for the different samples, the average exchange current density $\overline{i_{0,a/c}}$, given by Eq. (25) over the whole electrode thickness was kept constant:

$$\overline{i_{0,a/c}} = \frac{1}{L}\left(\int_{0}^{\omega L} i_{0,a/c}\left(dg_{a/c}^{0\ \text{to}\ \alpha L}\right) + \int_{\omega L}^{L} i_{0,a/c}\left(dg_{a/c}^{\alpha L\ \text{to}\ L}\right)\right)dx \tag{25}$$

Cathode				Anode			
% thickness ω	dg_{ω} [μm]	$dg_{1-\omega}$ [μm]	Quote	% thickness ω	dg_{ω} [μm]	$dg_{1-\omega}$ [μm]	Quote
20	2.0	4.048	Ic	20	2.0	4.048	Ia
20	1.8	6.654	IIc	20	1.8	6.654	IIa
10	2.0	3.324	IIIc	10	2.0	3.324	IIa
10	1.8	3.563	IVc	10	1.8	3.563	IVa
10	1.5	4.950	Vc	10	1.5	4.950	Va

Table 6.
Summary of the simulated microstructures.

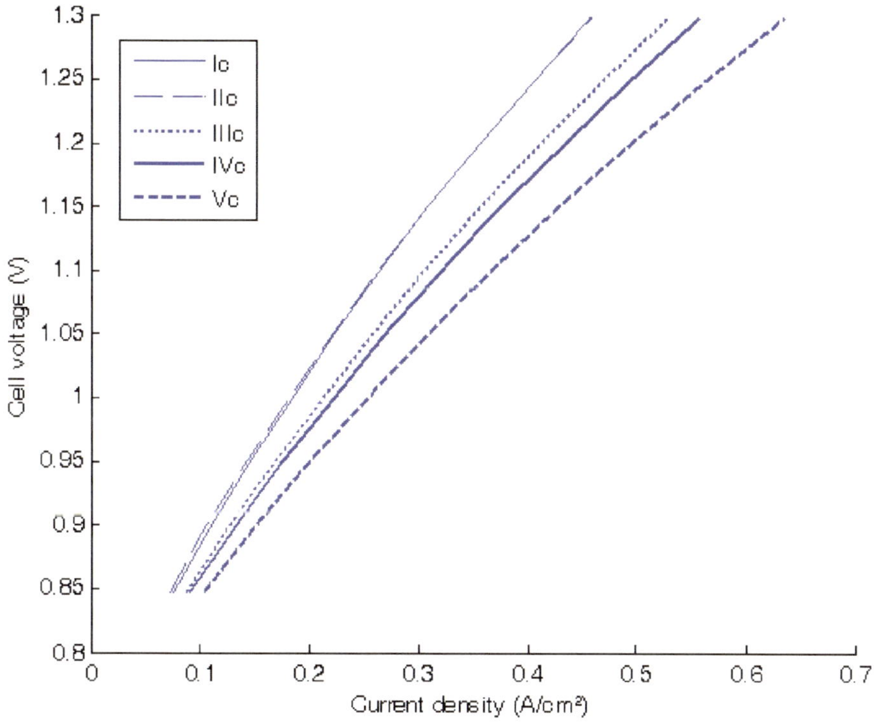

(a)

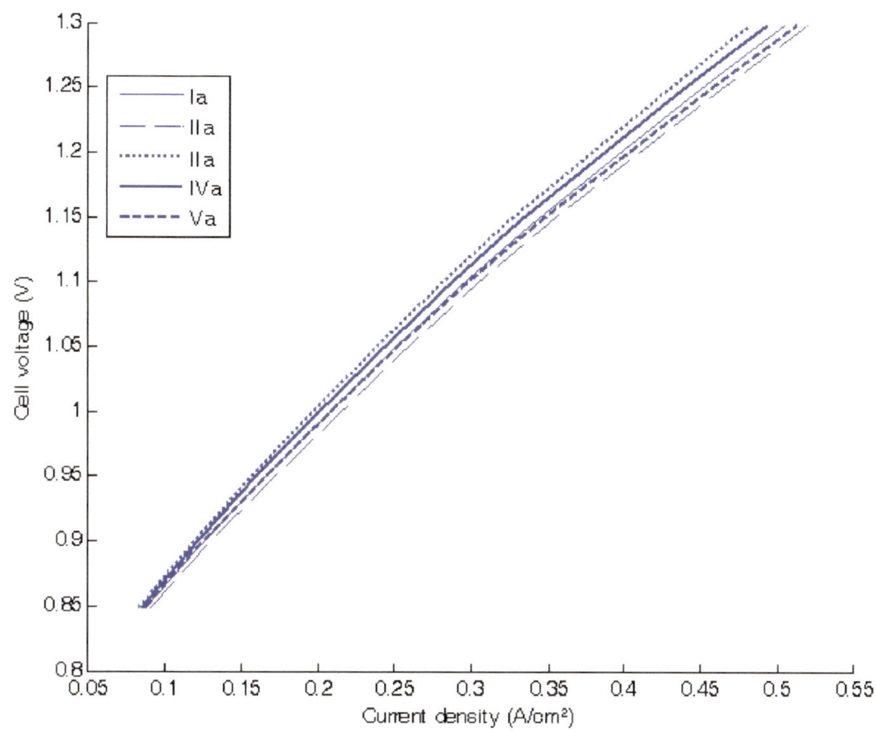

(b)

Figure 8.
Polarization curves for cathode simulations Ic to Vc (a) and anode simulations Ia to Va (b).

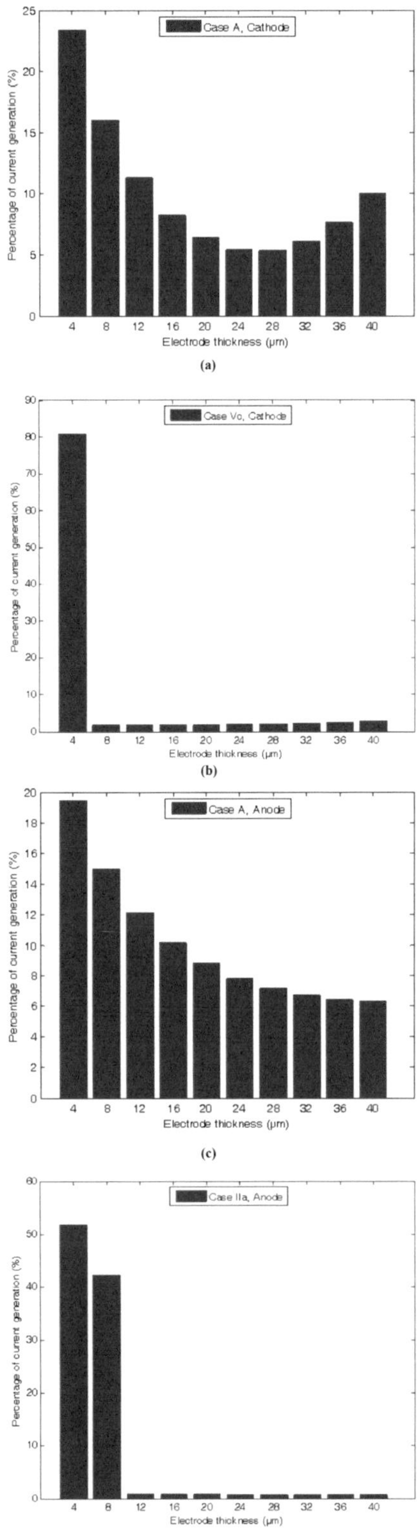

Figure 9.
Current sources distribution through both electrodes underneath the middle of the gas channel (x = 0 mm) for case A and cases displaying the best performance in the grain size investigation (Vc and IIa) at cell voltage = 1.3 V.

To do so in multilayer samples, the increase in exchange current density caused by a layer composed of smaller particles was compensated by a larger particle layer. The different grain diameters were adapted according to the thickness of the layer so that the average exchange current density remains constant. The layer displaying the thinner grains is always closer to the electrolyte and will be referred to as layer ω and expressed as a percentage of the total thickness of the electrode.

The current–voltage characteristics obtained for the five cathodic and anodic cases are displayed in **Figure 8a**, **b**. On the cathode side, sample Vc seems to present an obvious performance increase if compared to the other modified cathodes. Additionally, case IVc shows higher current densities than case IIc, even if both electrodes are composed of grain of equal diameter in the layer closest to the electrolyte. On the other hand, the microstructural changes do not influence significantly the electrical anode behavior. The polarization curves can barely be differentiated.

Ni et al. [15, 33] have evaluated the potential of functionally graded materials for SOEC electrodes. These authors have compared [33] conventional nongraded electrodes with particle size-graded electrodes. For graded electrodes with a particle size decreasing by 50% from the gas/electrode interface (dg = 1 μm) to the electrode/electrolyte interface (dg = 0.5 μm), a negligible reduction in potential is observed in comparison with nongraded electrode. For a higher gradient of particle size decreasing by 70% (i.e., dg = 1 ➔ 0.3 μm) a significant saving of H_2 electrode overvoltage has been observed. The present study exhibits similar results. In **Figure 8**, the bilayer with a 50% decrease in particle size (Ic: dg = 4.048 ➔ 2 μm) showed poor improvement of performance. In addition, a 70% grain diameter decrease (Vc: dg = 4.95 ➔ 1.5 μm) led to a large enhancement of the current generated by the cell.

Thin particles at the electrode/electrolyte interface (cathode interlayer thickness) decrease the area-specific resistance (ASR). This increase in performance goes along with the relocation of the electrochemical reaction at the electrode/electrolyte interfaces. As shown in **Figure 9**, 80% of the cathodic current arises from the first 10% of electrode thickness (case Vc), and on the anode side, more than 90% is generated by the first 20% (case IIa). The relocation in the volume observed for the smallest particles can easily be explained by the dependence of exchange current densities on grain diameter Eq. (10). Furthermore, forcing the current generation close to the electrolyte interface enables decreasing the ionic current path length.

5. Conclusion

The results show the influence of current collectors on gas access. Relevant control of material microstructures improves the diffusion of gaseous reactants and the current collecting. When diffusion is the limiting step, a relocation of the current sources within the volume of the electrodes is observed. On the contrary, if ionic ohmic drop becomes the rate determining step, current density sources are located close to the electrode/electrolyte interface. In some specific cases, the assumption that all electrochemical reactivity is located in the electrolyte interface vicinity cannot be made. That observation is emphasized when the competition between gas and ion transport is intentionally distorted, since a second reactive layer appears close to the cathodic gas channel.

It has been shown that it is possible to force the reaction to occur close to the electrolyte/electrode interfaces by layering the electrodes and introducing gradients of grain diameters. The obtained relocation is as high as 80% of the current being generated within the first 4 μm of the cathode thickness. The ohmic losses are reduced and gas access facilitated.

Contrary to the cathode, the changes of grain diameter gradient do not influence the electrochemical performance of the anode. However, its current production profile is consistently similar to the cathodic one. This means that specific attention should be given to the cathode microstructure.

Several aspects have been neglected in the present work and should be investigated to complete this approach and give global vision of the mechanisms that govern a SOEC response, e.g., the ohmic resistance due the dense ceramic membrane can be minimized using metal support technologies. In addition, the contact resistances shall be taken into consideration since they are a key parameter when optimizing the configuration of current collectors and will allow this model to be compared to experimental data of total SRU response.

Conflict of interest

No conflict of interest.

Acronyms and abbreviations

C	concentration [mol m^{-3}]
D	diffusion coefficient [m^2 s^{-1}]
dg	mean grain diameter [m]
dp	mean pore diameter [m]
E	Nernst potential [V]
i	current density [A m^{-3}]
L	electrode thickness [μm]
i_0	exchange current density [A m^{-3}]
M	molar mass [kg mol^{-1}]
n	normal derivative vector [m]
P	pressure [Pa]
Q	volumic current source [A m^{-3}]
T	temperature [K]
x	relative abscissa of electrode thickness [m]
V	voltage [V]
Y	molar fraction or volume fraction [−]

Greek

α	factor of symmetry [−]
β	fick diffusion form coefficient [−]
Γ	mass source term [$mol\ m^{-3}$ s^{-1}]
ε	porosity [−]
η	overpotential [V]
σ	conductivity [S m^{-1}]
τ	tortuosity [−]
ϕ	electrical potential [V]
Ω	ratio of electrode thickness [−]

Subscripts, superscripts

0	Standard
i	type of conductor

k	type of species
a	Anode
c	Cathode
a/c	anode or cathode
ω	axis position
S	electronic conductor
M	ionic conductor
S/M	electronic or ionic conductor
YSZ	yttria-stabilized zirconia
LSM	strontium-doped lanthanum manganite $La_{1-x}Sr_xMnO_3$
Ni	nickel
eff	effective
bv	Butler-Volmer
ref	reference
eq	equilibrium
pol	polarization
K	Knudsen
tot	total

Appendix

The diffusion coefficient for binary mixture of gases may be estimated from Fuller, Schettler, and Giddings relation with values coefficients for different molecules tabulated in [34]:

$$D_{H_2O\_H_2} = 3.16 \times 10^{-8} \frac{T^{1.75}\left(\frac{1000}{18} + \frac{1000}{2}\right)^{1/2}}{P\left(\left(7.06 \times 10^{-6}\right)^{1/3} + \left(12.7 \times 10^{-6}\right)^{1/3}\right)^2} \tag{26}$$

$$D_{O_2\_N_2} = 3.16 \times 10^{-8} \frac{T^{1.75}\left(\frac{1000}{28} + \frac{1000}{32}\right)^{1/2}}{P\left(\left(17.9 \times 10^{-6}\right)^{1/3} + \left(16.6 \times 10^{-6}\right)^{1/3}\right)^2} \tag{27}$$

Author details

Jonathan Deseure[1,2]* and Jérôme Aicart[3]*

1 LEPMI, CNRS, Grenoble Institute of Engineering (INP), University of Grenoble Alpes, Grenoble, France

2 University Savoie Mont Blanc, Grenoble, France

3 University Grenoble Alpes, CEA/LITEN, Grenoble, France

*Address all correspondence to: jonathan.deseure@univ-grenoble-alpes.fr and jerome.aicart@cea.fr

References

[1] Niether C et al. Improved water electrolysis using magnetic heating of FeC–Ni core–shell nanoparticles. Nature Energy. 2018;**3**(6):476-483

[2] Doenitz W, Schmidberger R, Steinheil E, Streicher R. Hydrogen production by high temperature electrolysis of water vapour☆. International Journal of Hydrogen Energy. 1980;**5**(1):55-63

[3] Erdle E, Donitz W, Schamm R, Koch A. Reversibility and polarization behaviour of high temperature solid oxide electrochemical cells. International Journal of Hydrogen Energy. oct. 1992;**17**(10):817-819

[4] Marina OA et al. Electrode performance in reversible solid oxide fuel cells. Journal of the Electrochemical Society. 2007;**154**(5):B452

[5] Zheng Y et al. A review of high temperature co-electrolysis of H_2O and CO_2 to produce sustainable fuels using solid oxide electrolysis cells (SOECs): Advanced materials and technology. Chemical Society Reviews. 2017;**46**(5): 1427-1463

[6] Kazempoor P, Braun RJ. Hydrogen and synthetic fuel production using high temperature solid oxide electrolysis cells (SOECs). International Journal of Hydrogen Energy. 2015;**40**(9): 3599-3612

[7] AlZahrani AA, Dincer I. Thermodynamic and electrochemical analyses of a solid oxide electrolyzer for hydrogen production. International Journal of Hydrogen Energy. 2017; **42**(33):21404-21413

[8] Jiang SP. Challenges in the development of reversible solid oxide cell technologies: A mini review: Challenges in the development of reversible SOC. Asia-Pacific Journal of Chemical Engineering. 2016;**11**(3): 386-391

[9] Cable TL, Setlock JA, Farmer SC, Eckel AJ. Regenerative performance of the NASA symmetrical solid oxide fuel cell design: Regenerative performance of the NASA symmetrical SOFC design. International Journal of Applied Ceramic Technology. 2011;**8**(1):1-12

[10] Graves C, Ebbesen SD, Jensen SH, Simonsen SB, Mogensen MB. Eliminating degradation in solid oxide electrochemical cells by reversible operation. Nature Materials. 2015;**14**(2): 239-244

[11] Laguna-Bercero MA. Recent advances in high temperature electrolysis using solid oxide fuel cells: A review. Journal of Power Sources. 2012; **203**:4-16

[12] Udagawa J, Aguiar P, Brandon NP. Hydrogen production through steam electrolysis: Model-based dynamic behaviour of a cathode-supported intermediate temperature solid oxide electrolysis cell. Journal of Power Sources. 2008;**180**(1):46-55

[13] Udagawa J, Aguiar P, Brandon NP. Hydrogen production through steam electrolysis: Control strategies for a cathode-supported intermediate temperature solid oxide electrolysis cell. Journal of Power Sources. 2008;**180**(1): 354-364

[14] Ni M, Leung MKH, Leung DYC. An electrochemical model of a solid oxide steam electrolyzer for hydrogen production. Chemical Engineering and Technology. 2006;**29**(5):636-642

[15] Ni M, Leung M, Leung D. Technological development of hydrogen production by solid oxide electrolyzer cell (SOEC). International Journal of

Hydrogen Energy. 2008;**33**(9): 2337-2354

[16] Laurencin J, Kane D, Delette G, Deseure J, Lefebvre-Joud F. Modelling of solid oxide steam electrolyser: Impact of the operating conditions on hydrogen production. Journal of Power Sources. 2011;**196**(4):2080-2093

[17] Grondin D, Deseure J, Ozil P, Chabriat J-P, Grondin-Perez B, Brisse A. Solid oxide electrolysis cell 3D simulation using artificial neural network for cathodic process description. Chemical Engineering Research and Design. 2013;**91**(1): 134-140

[18] Jin X, Xue X. Mathematical modeling analysis of regenerative solid oxide fuel cells in switching mode conditions. Journal of Power Sources. 2010;**195**(19):6652-6658

[19] Jin X, Xue X. Computational fluid dynamics analysis of solid oxide electrolysis cells with delaminations. International Journal of Hydrogen Energy. 2010;**35**(14):7321-7328

[20] Deseure J, Bultel Y, Dessemond L, Siebert E. Theoretical optimisation of a SOFC composite cathode. Electrochimica Acta. 2005;**50**(10): 2037-2046

[21] Deseure J, Bultel Y, Schneider LCR, Dessemond L, Martin C. Micromodeling of functionally graded SOFC cathodes. Journal of the Electrochemical Society. 2007;**154**(10):B1012

[22] Virkar A. The role of electrode microstructure on activation and concentration polarizations in solid oxide fuel cells. Solid State Ionics. 2000; **131**(1-2):189-198

[23] Kenney B, Karan K. Engineering of microstructure and design of a planar porous composite SOFC cathode: A numerical analysis. Solid State Ionics. 2007;**178**(3-4):297-306

[24] Grondin D, Deseure J, Brisse A, Zahid M, Ozil P. Simulation of a high temperature electrolyzer. Journal of Applied Electrochemistry. 2010;**40**(5): 933-941

[25] Grondin D, Deseure J, Ozil P, Chabriat J-P, Grondin-Perez B, Brisse A. Computing approach of cathodic process within solid oxide electrolysis cell: Experiments and continuum model validation. Journal of Power Sources. 2011;**196**(22):9561-9567

[26] Costamagna P. Modeling of solid oxide heat exchanger integrated stacks and simulation at high fuel utilization. Journal of the Electrochemical Society. 1998;**145**(11):3995

[27] Aguiar P, Adjiman CS, Brandon NP. Anode-supported intermediate temperature direct internal reforming solid oxide fuel cell. I: Model-based steady-state performance. Journal of Power Sources. 2004;**138**(1-2):120-136

[28] Achenbach E. Three-dimensional and time-dependent simulation of a planar solid oxide fuel cell stack. Journal of Power Sources. 1994;**49**(1-3):333-348

[29] Juhl M, Primdahl S, Manon C, Mogensen M. Performance/structure correlation for composite SOFC cathodes. Journal of Power Sources. 1996;**61**(1-2):173-181

[30] Hussain MM, Li X, Dincer I. A numerical investigation of modeling an SOFC electrode as two finite layers. International Journal of Hydrogen Energy. 2009;**34**(7):3134-3144

[31] Holtappels P, Bagger C. Fabrication and performance of advanced multi-layer SOFC cathodes. Journal of the European Ceramic Society. 2002;**22**(1): 41-48

[32] Haanappel VAC et al. Optimisation of processing and microstructural parameters of LSM cathodes to improve the electrochemical performance of anode-supported SOFCs. Journal of Power Sources. 2005;**141**(2):216-226

[33] Ni M, Leung MKH, Leung DYC. Mathematical modeling of the coupled transport and electrochemical reactions in solid oxide steam electrolyzer for hydrogen production. Electrochimica Acta. 2007;**52**(24):6707-6718

[34] Coulson JM, Richardson JF, Backhurst JR, Harker JH. Chemical Engineering. 4th ed. Oxford (England), New York: Pergamon Press; 1990

Chapter 7

A New Approach for Membrane Process Concentrate Management: Electrodialysis Bipolar Membrane Systems-A Short Communication

Taner Yonar

Abstract

In most cases traditional and advanced treatment technologies transfers and concentrates the pollutants from one phase to other phase. However, nowadays, these concentrated flows containing heavy pollution are rapidly moving away from being manageable. In particular, membrane concentrate flows await immediate solutions to this issue. Electrodialysis Bipolar Membrane (EDBM) Processes are becoming a serious and potential solution technique for similar concentrate streams. In this chapter, principles and potentials of EDBM processes for the recycling or recovery of membrane concentrates are discussed.

Keywords: electrodialysis, electrodialysis bipolar membrane, concentrate, reuse, recovery

1. Introduction

Membrane processes, including Ultrafiltration, Nanofiltration and Reverse Osmosis, are widely used for the treatment of water and wastewater. Main advantage of these processes are well known operational conditions and wide application areas. But, main disadvantage of these processes is concentrate management. Mainly high amount of salt content in concentrate stream is limiting the discharge of these streams to water bodies.

Starting from evaporation to ion exchange, most of these processes transfer the pollutants from one phase to other phase. It means more concentrated streams can be created from these processes. Therefore, applicable and valuable product production techniques are urgently needed for membrane concentrates. Electrodialysis Bipolar Membrane (EDBM) Processes are promising technique for the disposal of membrane concentrates.

EDBM technique has too many advantages against other techniques such as, valuable product formation and low cost operation. On the other hand, there is still great limitations on application point such as limited company production, alkali element fouling on membranes, high capital cost etc.,

Briefly, technological opportunities may solve the application disadvantages of EDBM processes in near future. In this chapter, principles and potentials of EDBM processes are discussed.

2. Electrodialysis bipolar membrane: principles and definitions

Industrial wastewater differs in industrial pollutant components depending on the types of industries in which they are formed. This difference plays a major role in the selection of wastewater treatment technologies.

Treatment technologies applied for wastewater recycling; Secondary Treatment (IA), Nutrient Removal, Filtration, Surface Filtration, Microfiltration (MF), Ultrafiltration (UF), Flotation, Nanofiltration (NF), Reverse Osmosis (TO), Electrodialysis (ED), Carbon Adsorption, Ion Exchange, Advanced Oxidation and Disinfection [1]. **Figure 1** shows the corresponding pore sizes of the pressure-operated membranes and their ability to hold specific wastewater components.

Membrane Processes (mainly reverse osmosis (RO) systems) and desalination plants are increasing day by day. In last two decades, over 10.000 membrane treatment plants have been established in most countries [3]. Daily treatment capacity of these plants may access 100 million m3/day in 2020.

With the introduction of low cost membrane modules in the 1960s and 1970s, membranes were widely used in industrial areas [4]. RO process (pore diameter < 0.0001 μm) can remove dissolved solids, bacteria, viruses and other microorganisms present in water [5]. By operating RO systems, which is one of the wastewater recycling processes, both high quality process water (filtrate flow) production is provided, and concentrate flow with high pollution load but low flow (silk) is formed. In RO systems with flow and concentrate modifications, approximately almost 90% filtrate and 10% concentrate can be formed from the inlet flow at high pressure [3].

The disposal of the concentrate from Membrane systems is still the main focus of most scientific research. This is an important issue for most country for the protection of water bodies and soil. As well known, discharged wastewater streams are still being used for irrigation. High salt content in concentrate streams means desertification of most valuable agricultural areas. The concentrate originating from membrane processes should be disposed or treated with feasible system.

Bipolar membranes are a type of ion-selective membranes first introduced in the 1950s (**Figure 2**). Bipolar membranes are composite membranes consisting of a Cation exchange membrane and an Anion exchange membrane [7]. Cation exchange membrane and anion exchange membranes, which are among ion exchange membranes, are heterogeneous, while bipolar membranes are homogeneous. Homogeneous bipolar membranes can be prepared from many different

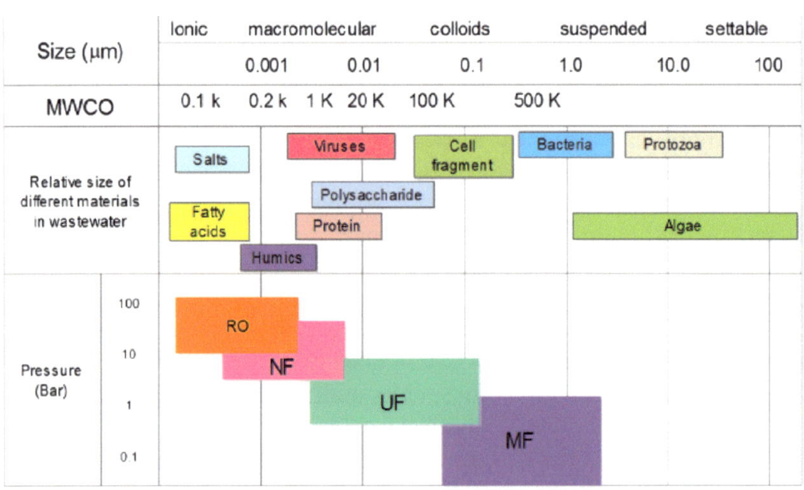

Figure 1.
Pressure operated membrane technology [2].

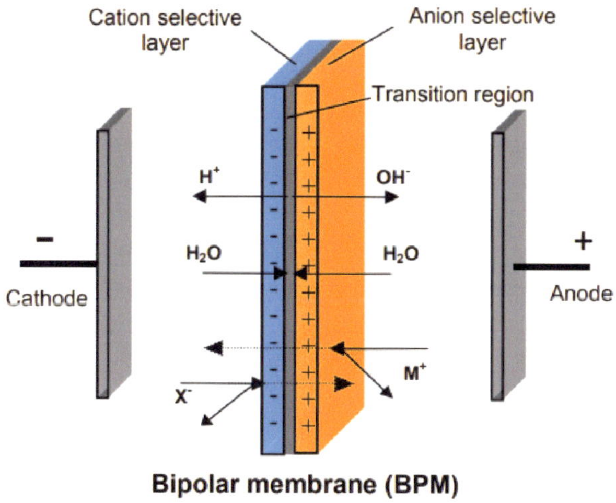

Figure 2.
Water splitting function of a bipolar membrane [6].

materials. The properties of bipolar membranes formed from different materials are given in the table below (**Table 1**).

Earlier studies on electrodialysis began before the World War II in Germany. Industrial and pilot scale applications have been developed since 1950. The first applications were about obtaining drinking water from sea water. Later, in applications in the food industry, it was tried to produce demineralized sugarcane sugar. If we consider the electrodialysis system as a black box, as a result of natural brine feeding, acid and base are formed at the output of the system, this picture is shown in the picture below [8] (**Figure 3**).

Electrodialysis processes is one of the membrane process that the driving force is an electrical field. EDBM system consist of anionic, cationic and bipolar membranes [9]. EDBM systems are widely used in chemical industry, in food industry, biochemistry industry and environmental protection technologies [9].

Using bipolar membranes together with ion exchange membranes in electrodialysis processes, high quality process water production can be possible and EDBM may become a more viable method for industrial wastewater treatment applications [10].

In the EDBM process, direct current (DC) is supplied to the electrodes to electrolyse the water molecules. Electrical potential between anode and cathode works as a driving force for the movement of electrons in electrolyte solution [8]. In EDBM process, bipolar membranes realizes the acid and base production in electrolyte solution.

Organic acids such as lactic acid, ascorbic acid, salicylic acid, amino acid and inorganic acids such as hydrofluoric (HF) acid, sulfuric acid (H_2SO_4), hydrochloric acid (HCl) can be produced using EDBM systems. Alkali bases potassium hydroxide (KOH), Sodium Methoxide (CH_3NaO) can also be produced in this systems [11].

In electrodialysis systems that separate water with bipolar membranes, an acid - base is formed from a very efficient energy-related salt concentration due to the accumulation of hundreds of cell units between 2 electrodes, such as conventional electrodialysis processes [7].

Some catalytic reactions take place in electrodialysis systems using bipolar membranes. Reactions between water molecules and functional chemical groups occur as Eq. (1), (2), (3), (4) [12]. The catalytic mechanism is underlined by chemical reaction model of water dissociation, that is, the water splitting could be considered as some type of proton-transfer reaction between water molecules and the functional groups or chemicals [7]:

Polymer(s)	Ion exchange group	Remarks
Anion exchange layer		
Poly-styrene-co- divinylbenzene	Tertiary and quaternary amines	Crosslinked resin, heterogeneous
Poly-sulfone	Di-amines	Crosslinked homogeneous
Poly-sulfone	Quaternary amines	Homogeneous
Poly-vinylidene fluoride blend with poly-vinyl benzyl chloride	Different diamines	Crosslinked
Anion exchange resin	Not specified	PVC binder
Poly-ether sulfone	Quaternary amines	Homogeneous
Poly-divinylbenzene-codimethylamino propyl methacrylamide		
Poly-methyl methacylate-coglycidyl methacrylate	Quaternary amines	
Cation exchange layer		
Poly-styrene-co- divinylbenzene	Sulfonic acid	Heterogeneous (polyvinylchloride binder)
Poly-styrene-co- divinylbenzene	Phosphoric acid	Poly-ethylene binder
Nafion	Sulfonic acid	Homogeneous
Grafted perfluorinated polymer membranes	Sulfonic acid	
Poly-butadiene-co-styrene	Sulfonic acid	
Poly-phenylene oxide or poly-styrene	Sulfonic acid	Homogeneous
Poly-ether sulfone	Sulfonic acid	Homogeneous
Poly-sulfone	Sulfonic acid	Homogeneous
Poly-ether ether ketone	Sulfonic acid	Homogeneous
Separate contact layer		
Poly-vinyl amine		
Poly-viylbenzylchrloride-co-divinylbenzene resin	Sulfonic acid	Heterogeneous (polyvinyl benzyl chlorideco-styrene binder)
Poly-sulfone	Aminated	

Table 1.
Ion exchange polymers of bipolar membrane layers [8].

$$B + H_2O \leftrightarrow BH^+ \ldots OH^- \leftrightarrow BH^+ + OH^- \tag{1}$$

$$BH^+ + H_2O \leftrightarrow B \ldots H_3O \leftrightarrow B + H_3O \tag{2}$$

$$A^- + H_2O \leftrightarrow AH \ldots OH^- \leftrightarrow AH + OH^- \tag{3}$$

$$AH + H_2O \leftrightarrow A^- \ldots H_3O^+ \leftrightarrow A^- + H_3O \tag{4}$$

where BH^+ and A^- refer to the catalytic centers. The catalytic sites provide an alternative path with low effective activation energy for water splitting into hydrogen and hydroxyl ions [7]. EDBM configurations including acid base production schematic diagrams are given in **Figure 4** [12].

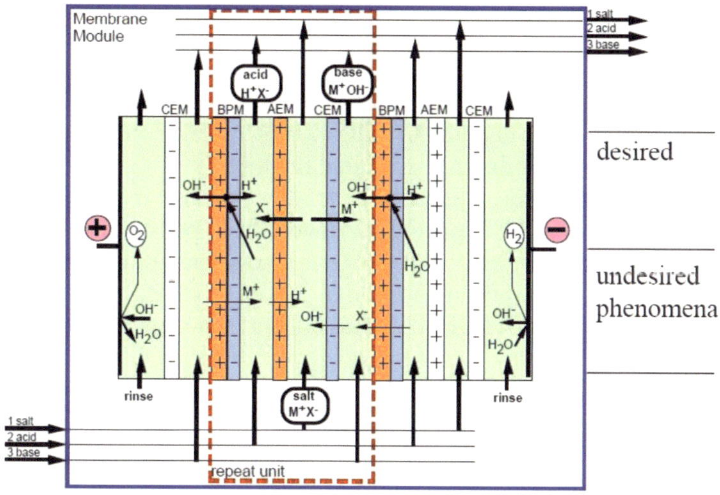

Figure 3.
Schematic illustration of EDBM process as a black-box [8].

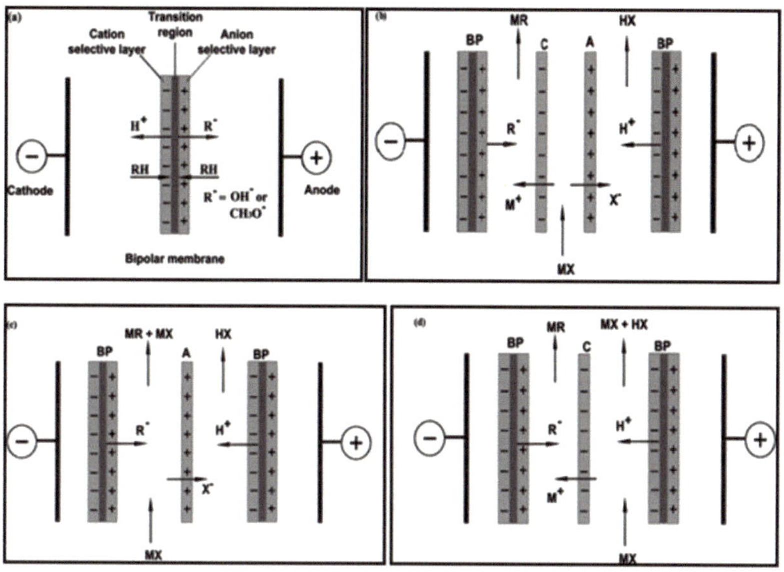

Figure 4.
Bipolar membrane and EDBM: BP, bipolar membrane; A, anion selective membrane; C, cation selective membrane; M +, cation; X- anion; H +, hydrogen ion; R, OH⁻ or CH_3O. (a) Bipolar membrane and its functions; (b) acid and base production; (c) acid production; (d) base production [12].

3. Usage areas of EDBM process

The bipolar membrane electrodialysis process is used in the latest technological way as an integrated process for the supply of potable water and industrial salt water and recovery of industrial effluent. On the other hand, both in chemical processes and environmental protection processes, they have been successfully applying in recent years. Another field of use of bipolar membrane is the chemical industry, where new products such as H_2SO_4 and NaOH are produced from a salt such as Na_2SO_4. Indeed, this method has become widespread recently and is used successfully. Especially, the production of acid and base without producing waste and the production of organic - inorganic acids increased the interest in this method. Many researchers have worked on this subject. It is possible to find many studies especially on acetic acid, propionic acid, gluconic acid, citric acid and lactic acid. In fact, some model studies have started to be carried out recently. Biotechnological research is ongoing to reduce costs in the electrodialysis process in order to reduce the cost in order to ensure acid recovery.

When ED and EDBM processes are compared with other treatment methods, it is an important advantage that it provides recovery from pollutants in wastewater and salt water. Studies show that it is possible to recover pollutants from solutions with one or more contaminants. In addition to this, the process of making acid and base production possible with EDBM process takes another step forward [13].

4. Recovery of concentrated wastes by EDBM process

An important advantage of the electrodialysis process over other processes is the possibility of recovery of concentrated waste. When the electrodialysis studies in the literature are taken into consideration, recovery has been proved possible. Electrodialysis studies are mainly in the form of recovery of those pollutants from aqueous solutions with single or multiple pollutants.

The process characteristics and economic evaluations of some studies in the literature are given in **Table 2**.

Application	Scale	Process characteristics	Economic evaluations	References
HF and HNO_3 Recovery	Industrial Scale	3 compartments, M. Area; $3x10^5$ m^2, BM Service Life; 2 years, Efficiency 90–95%	Total $ 2,950,000 Annual Business Administration Cost: $ 870,000	Pourcelly and Gavach [15]
NaOH Recovery	Semi Industrial Pilot Scale	M. Area: 0.5 m^2 Feed Speed 5 L / h Initial conc. 22 g / L Current Density 900 A/ m^2 Efficiency 82% (1 M)	5.0kWh/kg NaOH	[15]
NH_3 and HNO_3 Recovery GmbH, Germany	Semi Industrial Pilot Scale	M. Area: 120 m^2 Initial conc. 250 g / L Current Density 1000 A / m^2 Efficiency 97%	0.34 $/kg NaNO3	Pourcelly and Gavach [15], Graillon and Persin [16]
Dimethylisopropylamine recovery	Semi Industrial Pilot Scale	M. Area: 0,3 m^2 Initial conc 110 g / L Current Density 800 A / m^2 Efficiency 30–70%	$ 2.5–5.0 / kg amine	Pourcelly and Gavach [15], Graillon and Persin [16]
Gluconic Acid Recovery	Lab. scale	2 compartments, M. Area; 0,19 m^2, 2.2 V – 415 A/m^2 Efficiency 95%	—	Pourcelly and Gavach [15]
Methanesulfonic Acid Recovery	Industrial Scale	3 compartments, M. Area; 64 m^2, 2.26 V – 800 A/m^2 Initial conc 80–250 g / L Efficiency 95%	Total Expense: $ 700,000 Oper. Goods. $ 354 / ton MTA Sales fee: $ 5500 / ton MTA	Pourcelly and Gavach [15]

Application	Scale	Process characteristics	Economic evaluations	References
Amino Acid Recovery	Industrial Scale	3 compartments, M. Area; 540 m^2, BM Service Life; 2 years, Efficiency 4–6 M Org. Asit	—	Pourcelly and Gavach [15]
Lactic Acid Production	Industrial Scale	2 compartments, M. Area; 280 m^2, Efficiency 60–96%	UN Cost 0,12 $ / kg 1kWh / kg Acid	Pourcelly and Gavach [15], Aritomi [17]
Camphorsulfonic Acid Regeneration	Pilot Scale	3 compartments, BM. Area; 0,14m^2, Acid Conc. 0,8 M Current Density 500 A / m^2 Efficiency 98,5%	300 kWh/ton Acid	Pourcelly and Gavach [15]
Ascorbic acid Production	Lab. Scale - Semi End. Pilot Scale	2 Compartments Current Density 1000 A / m^2 Acid Conc. 1 M	1,4–2,3kWh/kg Acid	Pourcelly and Gavach [15], ve Novalic and Kulbe [18], Yu et al. [19]
Citric Acid Production	Pilot Scale	2 compartments, BM. Area; 0,004 m^2, Acid Conc. 30 M Current Density 1000 A / m^2	2–5 kWh/kg Acid	Wakamatsu [20], Xu [21], ve Novalic and Kulbe [18]
Salicylic Acid Production	Lab. - Pilot Scale	3 Compartments 30 V - 750 A / m^2 Acid Conc. 4.5 g / L	15–20 kWh / kg product	Alvarez et al. [22]
Sodium Acetate conversions	Pilot Scale	5 Compartments BM. Area; 0,008 m^2, Product 0.5 M Acetate	1.3–2.0 kWh / kg product	Yu et al. [19], Trivedi et al. [23]
Toluenesulphonic Acid regeneration	Lab. Scale	2 Compartments	1,2kWh / kg Acid	[19]
Formic Acid regeneration	Lab. Scale	3 compartments, M. Area; 540 m^2, Acid Conc. 7 M	2.6kWh / kg Acid	Ferrer and Laborie [24]
Sulfuric Acid recovery	Lab. Scale	6 compartments 3 compartments	3.3kWh / kg Acid 2.4kWh / kg Acid	Cifuentes [25]
Magnesium and Protein recovery	Lab. Scale	Bipolar Membrane electrodialysis	1.7kWh / kg Mg^{+2} 0.6kWh / kg protein	Pourcelly and Bazinet [26]

Table 2.
Recovery of concentrated wastes by EDBM process [14].

In addition to treatment and recovery, it is frequently encountered that electrodialysis method is used directly in acid and base production.

5. Advantages and disadvantages of EDBM process

Main advantage of EDBM systems is energy efficiency. In most cases high energy needed for most treatment processes mainly for pumping. But in EDBM process, system works with low pressure pumps (0.5–0.8 Bar). On the other hand direct current usage makes the EDBM systems advantageous against the other advanced treatment processes. Beside the most treatment processes, by products of EDBM are valuable materials such as acids and bases. Additionally, the inlet concentrated stream with a high salt content is deionized and water can be recovered. Briefly, salt content can be converted to valuable materials and water content can be reduced and recovered. Actually, it turns out that the EDBM system is a process capable of very high approach to zero waste practice.

The biggest advantages of using bipolar membranes in EDBM processes is that BPMs increase ionization by 50 million times compared to the self-hydrolysis of water. In addition, the anionic and cationic membranes inside the EDBM systems prevent the OH^- and H^+ ions formed in the system from reaching the anodes and cathodes, and no O_2 and H_2 gas output is observed in the BPMs [3].

The first disadvantage of EDBM systems meets the mark in acid and base production from membrane process concentrates. Concentrates consisting of membrane processes from wastewater treatment processes, generally contain mixed ion groups instead of single ion groups. This is due to the type and variety of wastewater they treat. Mixed acids and bases may remain low in commercial value. Another problem is the permanent damage left in the membrane structure of the multivalent ions in the concentrated stream. For this, +2 and + 3 valence cations can be removed from the water by nanofiltration, ion exchange, evaporation etc., before reverse osmosis process. Or + 2-valued (especially calcium compounds) ions can be removed from the reverse osmosis concentrate content by processes such as precipitation. However, in both cases, the operating costs of the processes will increase and will lead to reductions in acid and base concentrations resulting from the decrease in ion concentration in the feed water to EDBM systems.

Another problem to be encountered in EDBM systems may be the locking of the system at low acid and base concentrations (1–2%) with the increase of osmotic pressure in the system due to the increasing ion content. As it is known, high acid and base levels are important in these systems both commercially and in terms of water recovery. The most important solution related to this lies in the separate collection and purification of the salt content in the wastewater source. In other words, these systems can be paved with clean production. In other words, the disadvantages of the EDBM system in mixed wastewater streams can be prevented by interventions at the source. Thus, the operating life of EDBM systems is preserved and operating costs and product quality are increased.

It is certain that the advantages mentioned so far will lead the EDBM systems to attract more attention in the future. However, the biggest obstacle to the implementation of EDBM systems is the high cost of the membranes used in the system. This is thought to be in favor of the price decrease of the balances in the market due to the widespread use of EDBM and increased membrane production over time. Because the vast majority of existing wastewater treatment systems transfer existing pollutants to another phase or make them more concentrated and present them as an even bigger problem. However, EDBM systems promise us to use new products from our waste. This; It plays an important role in the solution of many environmental problems from efficient use of resources to global warming.

6. Conclusions

For further usage of membrane processes without any problematic concrete flows on environment, new concentrate disposal technologies will be needed such as EDBM process. The main advantage of the EDBM process is the commercially obtaining precious product from environmentally problematic products. However, the most important problem at the moment is high initial investment costs. But, similar high investment problems are valid for many processes that are widely used today, and with the spread of manufacturing, this problem has disappeared and the possibility of widespread use has increased. Finally, membrane concentrate flows are waiting an urgent solution, and the spread of EDBM or similar technologies will not take too long.

Author details

Taner Yonar
Environmental Engineering Department, Bursa Uludag University, 16059 Gorukle, Bursa, Turkey

*Address all correspondence to: yonar@uludag.edu.tr

References

[1] Visvanathan C, Asano T. The Potentıal for Industrıal Wastewater Reuse Environmental Engineering Program and Asian Institute of Technology. Department of Civil and Environmental Engineering, University of California; 2001. pp. 1-14

[2] Sadr SMK, Saroj DP. 14 - Membrane Technologies for Municipal Wastewater Treatment, In: Basile a, Cassano a, Rastogi N.K, Editors. Advances in Membrane Technologies for Water Treatment: Woodhead Publishing; 2015. pp. 443-446

[3] Badruzzaman M, Oppenheimer JS, Kumar M. Innovative beneficial reuse of reverse osmosis concentrate using bipolar membrane electrodialysis and electrochlorination processes. J. Membr. Science. 2009;**326**:392-399

[4] Adham S, Burbano A, Chiu K, Kumar M. Development of a NF/RO knowledgebase, California Energy Commission. In: Public Interest Energy Research Program Report. 2005

[5] Heijman SGJ, Guo H, Li S, van Dijk JC, Wessels LP. Zero liquid discharge: Heading for 99% recovery in nanofiltration and reverse osmosis. Desalination. 2009;**236**(1-3):357-362

[6] Balster J. Membrane Module and Process Development for Monopolar and Bipolar Membrane Electrodialysis, PhD Thesis. The Netherland: University of Twente; 2006

[7] Xu T. Ion exchange membranes: State of their development and perspective. Journal of Membrane Science. 2005;**263**:1-29

[8] Wilhelm FG. Bipolar Membrane Electrodialysis. PhD. Thesis. Enschede: Twente University; 2001

[9] Aksangür I. Investigation of Disposal and Optimization of Concentrate with Edbm System Which Originates from. Msc. Thesis. Bursa: Uludağ University; 2014

[10] Strathmann H. Electrodialysis, a mature technology with a multitude of new applications. Desalination. 2010;**264**:268-288

[11] Wilhelm FG, Punt IGM, Vegt NFA, Der V, Strathmann H, Wessling M. Cation permeable membranes from blends of sulfonated poly(ether ether ketone) and poly(ether sulfone). Journal of Membrane Science. 2002;**199**:167-176

[12] Yazıcı S. Analysis of Fouling Mechanism and Prevention Works of Electrodialysis with Bipolar Membrane Processes: Leachate Sample, Msc. Thesis. Istanbul: Yıldız Teknik University; 2012

[13] Yuzer B. Wastewater Treatment by Bipolar Membrane Electrodialysis Process and Evaluation of Reuse Alternatives. PhD. Thesis. Istanbul: İstanbul University-Cerrahpasa; 2018

[14] Ilhan F. Investigation of Treatability and Recycling of Landfill Leachate by Electrodialysis Process, PhD Thesis. Istanbul: Yıldız Teknik University; 2012

[15] Pourcelly, G. Gavach, C., (2000). Electrodialysis water splitting-application of electrodialysis with bipolar membranes, In Handbook on Bipolar Membrane Technology; Kemperman, A. J. B., Ed., Twente University Press: Enschede, The Netherlands, 17-46.

[16] Graillon, S.; Persin, F.; Pourcelly, G. ve Gavach, C., (1996). "Development of electrodialysis with bipolar membrane for the treatment of concentrated nitrate effluents", Desalination, 107: 159-169.

[17] Aritomi T., Nago S. ve Hanada F., (2001). "Performance of an improved bipolar membrane", Membrane Technology, 135: 11-13

[18] Novalic, S. ve Kulbe, K. D., (1998). "Separation and concentration of citric acid by means of electrodialytic bipolar membrane technology", Food Technology and Biotechnology, 36: 193-195.

[19] Yu, L. X., Guo, Q. F., Hao, J. H. ve Jiang, W. J., (2000). "Recovery of acetic acid from dilute wastewater by means of bipolar membrane electrodialysis", Desalination, 129: 283-288.

[20] Wakamatsu Y., Matsumoto H., Minigawa M. ve Tanioka A., (2006), "Effect of ionexchange nanofiber fabrics on water splitting in bipolar membrane", Journal of Colloid and Interface Science, 300(1): 442-445.

[21] Xu T. Electrodialysis processes with bipolar membranes (BMED) in environmental protection—A review. Resources Conservation Recycling. 2002;**37**(1)

[22] Alvarez, F., Alvarez, R., Coca, J., Sandeaux, J., Sandeaux, R. ve Gavach, C. (1997). "Salicylic acid production by electrodialysis with bipolar membranes", Journal of Membrane Science, 123: 61-69.

[23] Trivedi G.S., Shah B.G., Adhikary S.K., Indusekhar V.K. ve Rangarajan R., (1997). "Studies on bipolar membranes. part 11 – conversion of sodium acetate to acetic acid and sodium hydroxide", Reactive & Functional Polymers, 32: 209-215.

[24] Ferrer, J.S.J., Laborie, S., Durand, G. ve Rakib, M., (2006). "Formic acid regeneration by electromembrane process", Journal of Membran Science, 280(1-2): 509-516.

[25] Cifuentes, L., Garci'a, I., Ortiz, R. ve Casas, J. M., (2006). "The use of electrohydrolysis for the recovery of sulphuric acid from coppercontaining", Separation and Purification Technology, 50(2): 167-174.

[26] Pourcelly, G. Bazinet, L. (2007). In handbook of membrane separations: Chemical, pharmaceutical and biotechnological applications, Pabby, A. K., Rizvi, S. S. H., Sastre, A. M., Eds.; CRC Pr I Llc:, Florida.